AF307187

77 Springer Series in Solid-State Sciences

Edited by Peter Fulde

Springer Series in Solid-State Sciences

Editors: M. Cardona   P. Fulde   K. von Klitzing   H.-J. Queisser

J. Kondo   A. Yoshimori (Eds.)

Fermi Surface Effects

Proceedings of the Tsukuba Institute
Tsukuba Science City, Japan, August 27–29, 1987

With 76 Figures

Springer-Verlag Berlin Heidelberg New York
London Paris Tokyo

Dr. Jun Kondo

Electrotechnical Laboratory
Tsukuba, Ibaraki 305, Japan

Professor A. Yoshimori

Department of Material Physics, Osaka University,
Toyonaka-shi, Osaka 560, Japan

Series Editors:
Professor Dr., Dres. h. c. Manuel Cardona
Professor Dr., Dr. h. c. Peter Fulde
Professor Dr. Klaus von Klitzing
Professor Dr. Hans-Joachim Queisser

Max-Planck-Institut für Festkörperforschung, Heisenbergstrasse 1
D-7000 Stuttgart 80, Fed. Rep. of Germany

ISBN-13: 978-3-642-83427-1   e-ISBN-13: 978-3-642-83425-7
DOI: 10.1007/978-3-642-83425-7

2154/3150-543210

Preface

This volume is the proceedings of the Tsukuba Institute '87 on Fermi Surface Effects, which was held August 27–29, 1987, at Tsukuba Science City in Japan. The topic of the Institute, Fermi surface effects, is one of the fascinating subjects of solid-state physics. It has been known since Sommerfeld's work that the conduction electrons of metals constitute a degenerate Fermi system, and it has also been recognized that the occupation number of the electron states has a discontinuity across the Fermi surface. Several basic properties of metal electrons stem from this fact. Furthermore, it gives rise to a singular response of the metal electrons to local and dynamical perturbations of low frequency. Such singular behavior of the metal electrons is called a Fermi surface effect. In his opening address, printed as the Foreword, Professor R. Kubo described Fermi surface effects as due to "wild" behavior of the metal electrons.

The Institute consisted of five invited lectures, each of which was two hours long and dealt with theoretical aspects of a subject related to Fermi surface effects. Each lecturer is an expert in the field, and gave an intensive treatment of his own subject. The experiment of inviting only very few lecturers and allotting them ample time for both presentation and discussion seems to have been successful. This Institute, which was sponsored by the Japan Industrial Technology Association, will probably be followed by other institutes, forming a series.

The editors would like to express their sincere gratitude to the lecturers for presenting their talks and for preparing their manuscripts for the proceedings, to all the participants, and to the other members of the organizing committee of the Institute: Professor R. Kubo (chairman), Professor K. Yosida, Professor Y. Nagaoka, Professor A. Okiji and Professor K. Yamada. Special thanks are due to Dr. T. Ishiguro of the Electrotechnical Laboratory who, as secretary of the Institute, carried out much of the work necessary to organize this Institute and to make sure it ran smoothly.

Tsukuba, November 1987 *J. Kondo, A. Yoshimori*

Foreword

It is a great honor for me to make an address to welcome you to this Institute. It is a particularly great pleasure to meet here so many distinguished guests as well as young scientists from abroad and from within the country. This was made possible, in the first place, by the great efforts of the organizers, particularly Dr. Ishiguro and Dr. Kondo of the Electrotechnical Laboratory, and secondly, by the very appropriate planning to make this a sort of satellite conference of the 18th Low Temperature Physics Conference, which finished with great success last Wednesday.

This has become even more significant by the coincidence of the timing and the subject of the Institute with the award of the Fritz London Prize to Dr. Kondo at the LT Conference. The citation of the Prize reads "for his explanation of the phenomenon of the resistance minimum in metallic systems with magnetic impurities, revealing a subtlety in the behavior of interacting Fermi systems not suspected previously. Studies inspired by this work have led to major advances in theoretical techniques and to an improved understanding of a wide variety of interacting systems."

This coincidence is by no means just a coincidence, although surely it is not designed. The secrecy of the Prize was extremely secure. I did not know it until that morning. No one except the recipient knew that secret. But I would say that the coincidence was a sort of necessity. Both the award and the planning of this Institute were very appropriate. Therefore I ask the audience here to join me in presenting our hearty congratulations to Dr. Jun Kondo.

As you all know, the electron theory of metals was initiated by Drude and Lorenz at the beginning of this century. The Drude-Lorenz theory was on the right track but had inherent difficulties. A great breakthrough came in 1927 when Sommerfeld introduced the Fermi-Dirac statistics to the metallic electrons. This was soon followed by invention of band theories by Bloch and others. From this time, the Fermi sphere of free electrons and the Fermi surface in the momentum space became indispensable concepts of solid-state physics.

The totality of conduction electrons is sometimes called a Fermi sea. The surface is the Fermi surface. In the early years of solid-state physics, the Fermi sea had been thought as peaceful as the Pacific Ocean. Of course the surface is not rigid. There are tides and waves, surf, breakers and ripples. But they were thought mere perturbations. Gradually it has been realized that the Fermi sea is not really pacific. It is a dangerous sea. It is rather sensitive to outer disturbances. It reacts in complicated ways. It has many kinds of instabilities. Electrons are amiable creatures very familiar to us. They behave very nicely. We understand them very well. It is rather surprising to find electrons, when they form a big group, behave sometimes very wildly. In many cases their bad behavior is related to the existence of boundaries, namely the Fermi surface. Then it is called a Fermi surface effect. As you all know, in 1964 Dr. Kondo solved the problem of the resistance minimum, which had been discovered by the group at the Kamerlingh Onnes Laboratory in Leiden and remained an outstanding problem for nearly 30 years. Kondo's finding gave us a great incentive to deepen our understanding of metallic electrons interacting with various degrees of freedom in real systems. The purpose of this Institute is to provide a basis and an up-to-date review of theories on Fermi surface effects.

This Institute is sponsored by the Japan Industrial Technology Association, is supported by the Physical Society of Japan and the Agency of Industrial Science and Technology, which is a part of the world-famous MITI (Ministry of Trade and Industry), and is practically promoted by the Electrotechnical Laboratory. Nowadays there are a great number of scientific conferences and seminars all the time throughout the year and all over the world. In our country we have quite a number of such international physics meetings. Even so, this particular Institute is unique in one way. As far as I know, this is the first physics conference ever planned and organized by people at a government laboratory which is meant primarily to be technological. I do not want to dwell too much on this point, though I think it is an interesting phenomenon and I hope that this will be a good sign, showing that Japan is going to devote more effort to pursuing basic science, not only the technological developments.

Last but not least, I wish you all a pleasant and fruitful stay in Tsukuba.

Tsukuba   *Ryogo Kubo,*
August 27, 1987   *Chairman of the Organizing Committee*

Contents

Two-Level Systems in Metals

J. Kondo

Electrotechnical Laboratory, Tsukuba, Ibaraki 305, Japan

1. What is the Fermi Surface Effect?

As all of you know, the metal electrons constitute a degenerate Fermi system. Its excitation energy ranges from zero to several electron volts. Then, one may ask what is the energy scale of the metal electrons. In the case of static perturbations acting on the electrons, it is of the order of the Fermi energy ε_F . For example, suppose that an impurity potential $V(\mathbf{r})$ is placed in the jellium of the electrons. The energy shift due to this perturbation may be expanded in V:

$$\Delta E = c_1 V_0 + c_2 V_0^2 \rho + c_3 V_0^3 \rho^2 + \quad , \tag{1}$$

where V_0 is the matrix element of $V(\mathbf{r})$, which is assumed to be independent of the wave numbers, ρ is the density of the electron states. Thus, the expansion parameter is $V_0 \rho$, which is about V_0/ε_F . This means that the energy scale of the electrons is the Fermi energy. On the other hand, when the perturbation is dynamical and local, such as the s-d exchange model, excitation modes of low energy come into play and give rise to an infrared divergence. We call this fact the Fermi surface effect.

2. Brief Survey on Localized Spins in Metals

2.1 The s-d Model

As an introduction to the Fermi surface effect, I will make a brief survey of the s-d problem [1]. Here, we are concerned with magnetic impurities in metals, such as Mn in Cu, where the valence of the impurity is well defined. This system is represented by the s-d exchange model:

$$H_{sd} = \sum_{k\sigma} \varepsilon_k c_{k\sigma}^\dagger c_{k\sigma} - J \sum_{kk'} [(c_{k'\uparrow}^\dagger c_{k\uparrow} - c_{k'\downarrow}^\dagger c_{k\downarrow})S_z$$

$$+ c_{k'\uparrow}^\dagger c_{k\downarrow} S_- + c_{k'\downarrow}^\dagger c_{k\uparrow} S_+], \tag{2}$$

where $c^\dagger$ and c are the creation and annihilation operators for the electron and $\mathbf{S}$ is the spin operator for the impurity, which is assumed to have the magnitude 1/2 for simplicity.

When J=0, the ground state will be the Fermi sphere state Φ_0 multiplied by, say, the up-spin state α:

$$\Psi = \Phi_0 \cdot \alpha.$$

The first-order perturbation to the wave function due to J is expressed by

1

$$\Psi = \Phi_0 \cdot \alpha - J \sum_{\substack{k \leq k_F \\ k' \geq k_F}} \frac{1}{\varepsilon_k - \varepsilon_{k'}} \Phi_0(k\downarrow \to k'\uparrow) \cdot \beta + \cdots, \tag{3}$$

where an electron-hole pair ($k \to k'$) is created accompanied by spin-flip. The spin-non-flip term is of no interest for the present purpose. So, the z component of the spin decreases by perturbation. Its average by the wave function (3) is expressed by

$$\frac{\langle \Psi | S_z | \Psi \rangle}{\langle \Psi | \Psi \rangle} = \frac{1}{2} \frac{1 - J^2 \sum_{\substack{k \leq k_F \\ k' \geq k_F}} \frac{1}{(\varepsilon_k - \varepsilon_{k'})^2} + \cdot}{1 + J^2 \sum_{\substack{k \leq k_F \\ k' \geq k_F}} \frac{1}{(\varepsilon_k - \varepsilon_{k'})^2} + \cdot}. \tag{4}$$

The sum which occurs here is logarithmically divergent, because the excitation energy becomes as small as zero.

This example shows how a dynamical perturbation gives rise to the infrared divergence. In the case of a static perturbation, the normalization constant will also contain the same divergence. However, it will be cancelled by the same term in the numerator. Such cancellation does not occur in (4).

Now, on the basis of the s-d Hamiltonian (2), one can calculate the magnetic susceptibility, the specific heat or the electrical resistivity as a series of J. Such calculations were made by many authors (for reviews, see [1]), who found that each term of the series involved log T. It turned out that the actual expansion parameter is given by

$$\frac{J\rho}{1 - 2J\rho \, \log(kT/\varepsilon_F)}. \tag{5}$$

When J>0, this tends to zero as T→0. In this case, their theory gives a correct answer in the entire temperature range. When J<0, on the other hand, it diverges as T approaches T_K defined by

$$kT_K = \varepsilon_F \exp(1/2J\rho). \tag{6}$$

So, their theory was not correct below T_K. Equation (4) is an example of the difficulty at T=0. There has been an intensive effort to overcome this difficulty, and we now have even the exact solution of the problem. However, we will here digress from the s-d model and touch upon the Anderson model.

2.2 The Anderson Model

The Anderson model [2] describes the transition metal impurity in other metals. The subject was first studied by J. Friedel and the Anderson model extracts the essence of Friedel's theory. Anderson assumes that the transition metal impurity introduces extra orbitals of d-character which may be hybridized with the conducting states. The simplest Anderson model deals with an s-orbital instead of a d-orbital, and is described by

$$H_{AN} = \sum_{k\sigma} \varepsilon_k c_{k\sigma}^\dagger c_{k\sigma} + \varepsilon_d \sum_\sigma d_\sigma^\dagger d_\sigma + U d_\uparrow^\dagger d_\uparrow d_\downarrow^\dagger d_\downarrow$$
$$+ V_{sd} \sum_{k\sigma} (c_{k\sigma}^\dagger d_\sigma + d_\sigma^\dagger c_{k\sigma}). \tag{7}$$

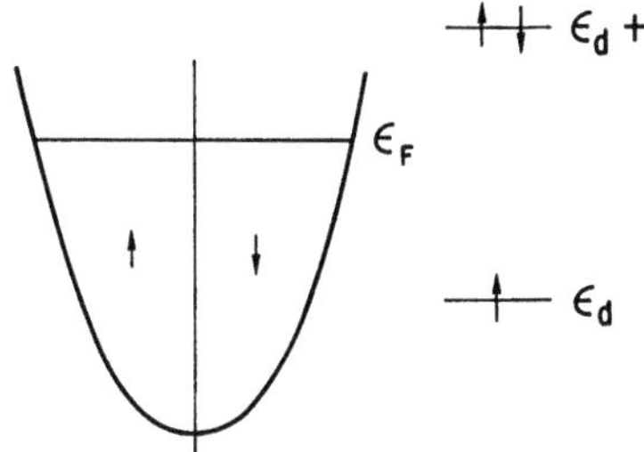

Fig. 1 Energy scheme of the Anderson model

Here, ϵ_d is the energy of the extra orbital, which we still call a d-orbital, U is the intra-orbital Coulomb repulsion and the last term represents hybridization.

We consider the case where ϵ_d is well below ϵ_F , so the d-state is always occupied by at least one electron. When it is occupied by the second elec-tron of the opposite spin, an extra energy U is necessary (Fig. 1). When the second level is well above the Fermi level, the valence of the d-state is almost fixed at one. This is the case where the impurity has a localized spin. In this case, the degrees of freedom for the d-state come from the directions of the spin and the Anderson model reduces to the s-d model, with the effective exchange interaction J given by

$$ J = - V_{sd}^2 \left(\frac{1}{\epsilon_F - \epsilon_d} + \frac{1}{U + \epsilon_d - \epsilon_F} \right) . \tag{8}$$

This means that J is intrinsically negative, so we encounter the same diffi-culty at low temperatures as in the case of the s-d model, if perturbation in V_{sd} is made.

2.3 The Yosida-Yamada Theory

YOSIDA and YAMADA [3] made a perturbation in U. The case of vanishing U is exactly soluble and if ϵ_d is also zero ($=\epsilon_F$), the d-state is occupied by electrons of both spins. When U is increased, the properties of the system may change continuously and reach the correct s-d limit for large U. In this case one will obtain correct answers by perturbation in U at least at low temperatures. This is the motivation of their study and they obtained several rigorous results, including the Wilson ratio being equal to two.
The picture of the local moment in metals obtained in this way is as follows: As U is increased with ϵ_d also changed according to $\epsilon_d = -U/2$ (symmetric case), the d-state is occupied by electrons with an unequal number of up and down spins. The direction of the major spin is fluctuating in time (Fig. 2). On the average, the occupation numbers of both directions of the spin are equal. However, as U is increased, the fluctuation time is also increased. If the time necessary to make an observation is short compared with the fluctuation time, one observes quantities related to instantaneous directions of the spin. For thermodynamic problems, the observation time is about $\hbar/kT$, so it is short at high temperatures. More precisely, when T is larger than T_K, the fluctuation time is longer than the observation time. In this case, the model of the local spin is appropriate. In the converse case, the local moment is averaged out and perturbation expansion starting from the state with vanishing U is appropriate.

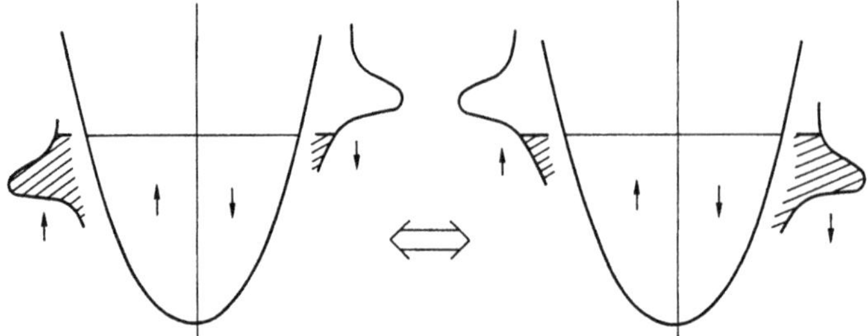

Fig. 2   The Anderson model at finite temperatures

2.4 The Wilson Theory

The Yosida-Yamada theory is applicable at low temperatures. Now, WILSON [4]
developed a theory of the s-d model using a renormalization group method.
With the help of numerical computation, he obtained results which are valid
in the whole temperature range. His solution at high temperatures is
essentially identical to that of the previous authors, whereas that at low
temperatures is identical to the Yosida-Yamada theory. Thus, he was able to
correlate parameters which occur in high temperature expansion with those
which occur in the low temperature theory. Thus, he solved the problem in
the entire temperature range.

2.5 The Exact Solution of the s-d Problem

Recently, TSVELICK and WIEGMANN [5] and ANDREI, FURUYA and LOWENSTEIN [6]
obtained exact solutions of the s-d problem by reducing the model to that of
a one-dimensional electron gas interacting with a local spin. They obtained
results which agreed with those of Wilson and also several new results.
Their method can be extended to a more general s-d problem, and also to the
Anderson model. In the latter case, one can treat the case where the valence
of the d-state is not fixed at one but may fluctuate. These subjects will be
discussed by Professor Okiji.

3. Brief Survey of the X-Ray Spectra of Metals

3.1 The Mahan Theory

The infrared divergence associated with soft X-ray absorption or emission of
metals was first studied by MAHAN [7]. Suppose that an inner-shell electron
is raised to a conducting state **p**. The potential of the hole left in the
inner shell will scatter the electron to another conducting state **k**. This
scattering cannot be considered as a single-electron problem, because the
electron cannot be scattered into the filled states. So, the wave function
after the photon absorption will read

$$\left(a_{\mathbf{p}}^{\dagger} + \sum_{k \geq k_F} \frac{V_0}{\varepsilon_{\mathbf{p}} - \varepsilon_{\mathbf{k}}} a_{\mathbf{k}}^{\dagger} + \cdots \right) \cdot \Phi_0 \,, \qquad (9)$$

where the k sum extends over the states above the Fermi level. The absorption intensity is given by

4

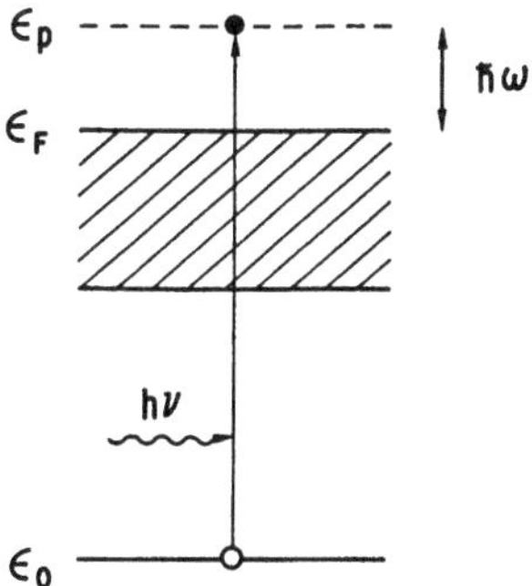

Fig. 3 Soft X-ray absorption of metals

$$I(\omega) \propto \sum_p \left(1 + \sum_{k \geq k_F} \frac{V_0}{\epsilon_p - \epsilon_k} + \cdot\cdot \right)^2 \delta(\hbar\omega - (\epsilon_p - \epsilon_F)) \; , \tag{10}$$

where $\hbar\omega = \hbar\nu - (\epsilon_F - \epsilon_0)$ is the photon energy measured from the Fermi level (Fig. 3). Since the state k is restricted to those above the Fermi level, the sum in (10) becomes the logarithm of the electron energy:

$$\sum_{k \geq k_F} \frac{V_0}{\epsilon_p - \epsilon_k} = V_0 \rho \; \log \frac{|\epsilon_p - \epsilon_F|}{\epsilon_F} \; .$$

Using this result in (10), one finds $I(\omega) \propto 1 + 2V_0 \rho \cdot \log (\hbar\omega/\epsilon_F) + \cdot\cdot$. Calculating higher-order terms, Mahan found that the series is summed to become

$$I(\omega) \propto \left(\frac{\hbar\omega}{\epsilon_F} \right)^{2V_0\rho} . \tag{11}$$

Since the hole potential is attractive, (11) diverges as $\omega \to 0$.

3.2 The Nozières-de Dominicis Theory

The hole in the inner shell will also scatter the electrons of the Fermi sea and produce many electron-hole pairs. This fact is also important in the X-ray spectra, and this problem was studied by NOZIÈRES and DE DOMINICIS and collaborators [8].

Now, forget about the electron raised from the inner shell and just assume that the hole potential was suddenly switched on at t=0. Then, the wave function of the Fermi sea at a later time t is expressed by

$$\Phi(t) = e^{-iHt/\hbar} \, \Phi_0 \; , \tag{12}$$

where H includes the hole potential:

$$H = H_0 + \sum_i V(\mathbf{r}_i) . \tag{13}$$

They showed that the absorption intensity involves

$$I'(\omega) = \int \left| \langle \Phi(t) | \Phi_0 \rangle \right|^2 e^{-i\omega t} \, dt$$

as a factor. For large t, $\Phi(t)$ involves many electron-hole pairs of low excitation energy. They found that at $t \gg \hbar/\varepsilon_F$

$$\langle \Phi(t) | \Phi_0 \rangle = \exp\left[-\left(\frac{\delta}{\pi}\right)^2 \log(\varepsilon_F t/\hbar)\right] = \left(\frac{\hbar}{\varepsilon_F t}\right)^{\left(\frac{\delta}{\pi}\right)^2} , \tag{14}$$

where δ is the s-wave phase shift for the hole potential, which is assumed to cause only s-wave scattering. Then, $I'(\omega)$ is calculated as

$$I'(\omega) = \left(\frac{\hbar\omega}{\varepsilon_F}\right)^{2\left(\frac{\delta}{\pi}\right)^2} .$$

The total intensity is now expressed by

$$I(\omega) \propto \left(\frac{\hbar\omega}{\varepsilon_F}\right)^{\left[-\frac{2\delta}{\pi} + 2\left(\frac{\delta}{\pi}\right)^2\right]} , \tag{15}$$

where the first term in the exponent corresponds to what Mahan found. Equation (15) diverges or vanishes depending on whether the exponent is negative or positive. This fact is called edge singularity and will be discussed by Professor Mahan in detail.

4. Anderson's Orthogonality Theorem

We note that (14) tends to zero as time goes to infinity. This fact is usually referred to as Anderson's orthogonality theorem. ANDERSON [9] considered a static problem, i.e., he considered the overlap integral between the ground state of H_0 and that of H and found that the integral is expressed by

$$\exp\left[-\left(\frac{\delta}{\pi}\right)^2 \log(\varepsilon_F/\Delta\varepsilon)\right] , \tag{16}$$

where δ is the s-wave phase shift for $V(\mathbf{r})$ and $\Delta\varepsilon$ is the level separation between s-states at the Fermi level. Since $\Delta\varepsilon$ is inversely proportional to the system size, (16) tends to zero as the system size tends to infinity.

I will mention a version of the theorem, which is relevant to my later talk. Let Φ_1 be the ground state obtained when the impurity is fixed at $\mathbf{R}_1$:

$$H_1\Phi_1 = E\Phi_1 ,$$

$$H_1 = H_0 + \sum_i V(\mathbf{r}_i - \mathbf{R}_1). \tag{17}$$

Similarly, Φ_2 and H_2 are defined for the impurity fixed at $\mathbf{R}_2$. Then, we are concerned with the overlap integral $\langle \Phi_1 | \Phi_2 \rangle$, which is expressed by

$$\langle \Phi_1 | \Phi_2 \rangle = \exp\left[-K \cdot \log(\varepsilon_F/\Delta\varepsilon)\right] , \tag{18}$$

where K is a function of V and the distance between the two positions. More

precisely, K is a function of phase shifts and the distance, and should
vanish when the distance vanishes. Even when V is spherically symmetric, our
problem lacks spherical symmetry and calculation of K is a rather difficult
task. Much effort has been applied by Yamada, Yosida and collaborators [10]
to this problem. They are now able to obtain the values of K in a general
case. We quote here the result of a simple case, where only the s-wave phase
shift is non-zero:

$$K = \frac{2}{\pi^2} \left[\tan^{-1} \left(\frac{\sqrt{1 - x \tan\delta}}{\sqrt{1 + x\tan\delta}} \right) \right]^2 \leq \frac{1}{2} , \qquad (19)$$

$$x = \frac{\sin^2 k_F a}{k_F^2 a^2} , \qquad a = |\mathbf{R}_1 - \mathbf{R}_2| ,$$

$$\tan\delta = -\pi V_0 \rho .$$

When $V_0 \rho$ is small, K is approximated by

$$K = 2V_0^2 \rho^2 \left(1 - \frac{\sin^2 k_F a}{k_F^2 a^2} \right) . \qquad (20)$$

This is what I obtained in discussion of a two-level system in metals [11],
and it will often be referred to in a later part of this talk. When the
distance a is small, K is proportional to the square of a:

$$K \approx \frac{2}{3\pi^2} \frac{\tan^2\delta}{1 + \tan\delta} k_F^2 a^2 . \qquad (21)$$

This result will also be used later.

5. Non-adiabatic Effect for Heavy Particles in Metals

Let us consider protons or muons in metals. When Nb, for example, is exposed
to hydrogen gas, the H_2 molecules become attached to the metal surface and
dissociate into H atoms, which then diffuse into the body of the metal.

Positive muons, after injection into metals, are believed to lose their
kinetic energy immediately and diffuse with thermal energy before they decay
with the life time of 2.2μs. They are regarded as an isotope of the proton.
Both particles sit in interstitial sites in metals and diffuse between them.
At high temperatures, their diffusion constant is of activation type and
decreases rapidly as the temperature goes down. At low temperatures, they
may diffuse by tunneling between sites. We are interested in the effect of
metal electrons on such tunneling.

5.1 Two-Site Model

When tunneling motion is slow, we may consider only two sites, because
successive tunnelings will not be correlated. Assume the potential U($\mathbf{R}$) for
the particle consists of two wells, whose centers are at $\mathbf{R}_1$ and $\mathbf{R}_2$. Let the
lowest orbit in each well be χ_1 or χ_2. Then, the eigenstates of the particle
will be expressed by

$$\chi(\mathbf{R}) = \frac{1}{\sqrt{2}}\,(\chi_1 \pm \chi_2)\;, \tag{22}$$

with the energy given by

$$E = \pm\Delta,$$

$$\Delta = \int \chi_1(\mathbf{R})\; H'\; \chi_2(\mathbf{R})\; d^3\mathbf{R}. \tag{23}$$

Here, Δ is the tunneling matrix element between the two orbits.

We now consider the interaction between the particle and the electrons, so that the total Hamiltonian of our system reads

$$H = -\frac{\hbar^2}{2M}\,\Delta_{\mathbf{R}} + U(\mathbf{R}) + H_0 + V_{\mathbf{R}}\;, \tag{24}$$

$$H_0 = -\frac{\hbar^2}{2m}\,\Sigma_i\,\Delta_i\;,$$

$$V_{\mathbf{R}} = \Sigma_i\,V(\mathbf{r}_i - \mathbf{R}).$$

Here, the first term is the kinetic energy of the particle.

5.2 Adiabatic Approximation [12]

Since the particle mass is much larger than the electron mass, one might treat this problem in the adiabatic approximation. Then, one would first fix the particle at $\mathbf{R}$ and solve for the electron wave function:

$$(H_0 + V_{\mathbf{R}})\cdot\Phi_{\mathbf{R}}(\mathbf{r}_1,\,\mathbf{r}_2\cdot\cdot) = E_{\mathbf{R}}\Phi_{\mathbf{R}}(\mathbf{r}_1,\,\mathbf{r}_2\cdot\cdot)\;, \tag{25}$$

where $E_{\mathbf{R}}$ is the adiabatic potential for the particle. (In our jellium model, it is a constant.) Then, one would solve for the particle wave function:

$$\left(-\frac{\hbar^2}{2M}\,\Delta_{\mathbf{R}} + U(\mathbf{R}) + E_{\mathbf{R}}\right)\cdot\chi(\mathbf{R}) = E_{adi}\chi(\mathbf{R}). \tag{26}$$

The total wave function would be the product of the two:

$$\Psi = \chi(\mathbf{R})\cdot\Phi_{\mathbf{R}}(\mathbf{r}_1,\,\mathbf{r}_2\cdot\cdot)$$

$$= \frac{1}{\sqrt{2}}\,[\chi_1(\mathbf{R}) + \chi_2(\mathbf{R})]\cdot\Phi_{\mathbf{R}}(\mathbf{r}_1,\,\mathbf{r}_2\cdot\cdot). \tag{27}$$

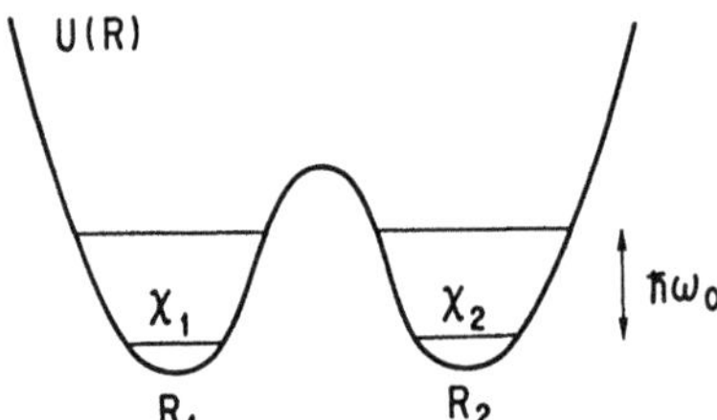

Fig. 4 Two-well potential for the particle

In this scheme, the effective tunneling matrix is the same as that of the bare particle, because the integration with respect to the electron coordinates gives us unity:

$$\Delta_{eff} = \int \chi_1^*(\mathbf{R}) \cdot \phi_{\mathbf{R}}^*(\mathbf{r}_1, \cdot\cdot) \, H' \, \chi_2(\mathbf{R}) \cdot \phi_{\mathbf{R}}(\mathbf{r}_1, \cdot\cdot) \, d^3R \, d^3r_1 \cdot\cdot$$

$$= \int \chi_1^*(\mathbf{R}) \, H' \, \chi_2(\mathbf{R}) \, d^3R = \Delta. \tag{28}$$

In general, the adiabatic scheme is valid when the energy scale of the electrons is larger than that of the particle. However, this is not true in our case, because of low excitation modes of the electrons. Thus, the adiabatic scheme breaks down here. In fact, the average of the energy by Ψ , which is written as

$$\langle\Psi|H|\Psi\rangle = E_{adi} + \langle\chi(\mathbf{R})|\int - \frac{\hbar^2}{2M} \phi_{\mathbf{R}}^*(\mathbf{r}_1, \cdot\cdot) \, \Delta_{\mathbf{R}} \, \phi_{\mathbf{R}}(\mathbf{r}_1, \cdot\cdot) d^3r_1 \cdot\cdot |\chi(\mathbf{R})\rangle,$$

turns out to diverge, because the integral in the above expression, which is expressed as the second derivative of the overlap integral, is logarithmically divergent:

$$\int \phi_{\mathbf{R}}^*(\mathbf{r}_1, \cdot\cdot) \, \Delta_{\mathbf{R}} \, \phi_{\mathbf{R}}(\mathbf{r}_1, \cdot\cdot) d^3r_1 \cdot\cdot = \Delta_{\mathbf{R}} \langle\phi_{\mathbf{R}'}|\phi_{\mathbf{R}}\rangle|_{\mathbf{R}'=\mathbf{R}} \propto -\log \frac{\epsilon_F}{\Delta\epsilon}.$$

Here, use has been made of (18) and (21). One must, therefore, proceed beyond the adiabatic approximation in some way, and then one will have a tunneling matrix different from the bare one.

5.3 Simple Non-adiabatic Scheme [12]

The simplest way to take account of the non-adiabatic effect will be to take the wave function as

$$\Psi = \frac{1}{\sqrt{2}} \, [\chi_1(\mathbf{R}) \cdot \phi_{\mathbf{R}_1}(\mathbf{r}_1, \cdot) \pm \chi_2(\mathbf{R}) \cdot \phi_{\mathbf{R}_2}(\mathbf{r}_1, \cdot\cdot)]. \tag{29}$$

For this wave function, the tunneling matrix is renormalized by the overlap integral between $\phi_{\mathbf{R}_1}$ and $\phi_{\mathbf{R}_2}$:

$$\Delta_{eff} = \Delta \cdot \langle\phi_{\mathbf{R}_1}|\phi_{\mathbf{R}_2}\rangle = \Delta \cdot \left(\frac{\Delta\epsilon}{\epsilon_F}\right)^K. \qquad (T=0) \tag{30}$$

So, Δ_{eff} vanishes as the system size tends to infinity. We will see later that, at finite temperatures, the effective overlap integral is expressed by [11]

$$\Delta_{eff} = \Delta \cdot \left(\frac{kT}{\epsilon_F}\right)^K. \tag{31}$$

At ordinary temperatures, this renormalization factor is very small compared with unity and depends on the temperature, so the non-adiabatic effect in our case is dramatic. Although (29) is too simple, later study will show that the result (31) is essentially correct and relevant to some physical systems. The purpose of this talk is to discuss dynamics and thermodynamics of the particle in a two-well potential in the context of the non-adiabatic effect expressed by (31).

6. Partition Function of a Two-Well System

We consider the system represented by the Hamiltonian (24) and will calculate the partition function:

$$Z = \text{Tr}\{e^{-\beta H}\}. \qquad (\beta = 1/kT) \qquad (32)$$

6.1 Feynman's Path Integral

We resort to the method of Feynman's path integral, which has recently been applied to the present system by GUINEA [13] and HEDEGARD [14] for dynamical problems. We first transform the trace over the particle coordinate into the integral over particle paths. Consider a path of the particle $R(u)$ for the time u between 0 and $\beta\hbar$, which satisfies the boundary condition

$$R(0) = R(\beta\hbar). \qquad (33)$$

Then, calculate the integral S_T:

$$S_T = \int_0^{\beta\hbar} \left[\frac{M}{2}\left(\frac{dR}{du}\right)^2 + U(R(u)) + \sum_i V(r_i - R(u)) \right] du, \qquad (34)$$

which is a functional of the path. (Here, r_i may be considered as a c-number, since we may also apply the path-integral method to the traces over the electron coordinates.) Feynman showed that the trace over the particle coordinate is obtained by the path integral

$$\text{Tr}_R\{e^{-\beta H}\} = \int \cdots \int DR(u) \, \exp\left(-\beta H_0 - \frac{S_T}{\hbar}\right) , \qquad (35)$$

where $DR(u)$ denotes summing over all the particle paths that satisfy (33). Hereafter, we set $\hbar=1$.

6.2 Trace over Electron Coordinates

Next, we perform traces over electron coordinates. We are concerned with Z_e:

$$Z_e = \text{Tr}_e \left\{ \exp\left[-\beta H_0 - \int_0^\beta \sum_i V(r_i - R(u)) du \right] \right\} \qquad (36)$$

and expand it in terms of V:

$$Z_e = Z_{e0} \left(1 + \sum_n (-1)^n \int \cdots \int_{\beta > u_1 > \cdots > u_n > 0} \langle V(u_1) \cdots V(u_n) \rangle du_1 \cdots du_n \right) . \qquad (37)$$

Here Z_{e0} is the unperturbed electron partition function, and $V(u)$ is defined by

$$V(u) = e^{uH_0} \sum_i V(r_i - R(u)) \, e^{-uH_0}. \qquad (38)$$

Note that u is involved in $R(u)$, too.

It is more convenient to use the second quantization scheme to calculate the average in (37). In this scheme, we have

$$V(u) = V_0 \sum_{kk'\sigma} c_{k'\sigma}^\dagger c_{k\sigma} \, e^{i(k-k')\cdot R(u)} \, e^{(\varepsilon_{k'} - \varepsilon_k)u} . \qquad (39)$$

10

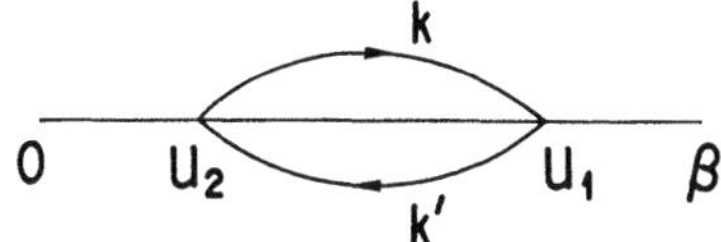

Fig. 5   The lowest-order bubble diagram

The lowest-order term (n=2) is represented by the diagram of Fig. 5 and is calculated as

$$\langle V(u_1)V(u_2)\rangle = 2V_0^2 \sum_{\mathbf{kk'}} e^{i(\mathbf{k-k'})\cdot[\mathbf{R}(u_1)-\mathbf{R}(u_2)]}$$
$$\times e^{(\varepsilon_{k'}-\varepsilon_k)(u_1-u_2)} f_{k'}(1-f_k)$$
$$= G(u_1, u_2), \tag{40}$$

where $f_{\mathbf{k}}$ is the Fermi distribution function. In the nth order terms, we take   only those diagrams where there are n/2 bubbles (Fig. 6). They give terms which involve the largest number of logarithmic factors. All those diagrams can be summed to give

$$Z_e = Z_{e0} \exp\left(\int_0^\beta \int_0^{u_1} G(u_1,u_2)du_1 du_2\right).$$

Then, the entire partition function is expressed as

$$Z = Z_{e0} \int\cdots\int D\mathbf{R}(u)\, e^{-S-S_I}, \quad \text{where} \tag{41}$$

$$S = \int_0^\beta \left[\frac{M}{2}\left(\frac{d\mathbf{R}}{du}\right)^2 + U(\mathbf{R}(u))\right] du, \tag{42}$$

$$S_I = -\iint_{\beta>u_1>u_2>0} G(u_1, u_2)\, du_1 du_2. \tag{43}$$

S is the action of the bare particle, whereas all the effects of the electrons are included in S_I.

We proceed to calculation of G, (40). We first perform the angular integrals for $\mathbf{k}$ and $\mathbf{k'}$ to find

$$G(u_1, u_2) = 2V_0^2 \sum_{\mathbf{kk'}} \frac{\sin k\Delta R}{k\Delta R}\cdot\frac{\sin k'\Delta R}{k'\Delta R}\, e^{(\varepsilon_{k'}-\varepsilon_k)(u_1-u_2)} f_{k'}(1-f_k),$$

where

$$\Delta R = |\mathbf{R}(u_1) - \mathbf{R}(u_2)|. \tag{44}$$

Fig. 6   Some of the most divergent diagrams

We may set k and k' in the above expression to k_F, then we find

$$G(u_1, u_2) = 2V_0^2 \frac{\sin^2 k_F \Delta R}{k_F^2 \Delta R^2} \sum_{k'} e^{\varepsilon_{k'}(u_1-u_2)} f_{k'} \sum_k e^{-\varepsilon_k(u_1-u_2)}(1-f_k). \tag{45}$$

Both sums in the above expression have the same value, if the energy band is symmetrical about the Fermi level. Assuming a constant density of states ρ , the sum is calculated as

$$\sum_k \exp(\varepsilon_k u) f_k = \frac{\pi\rho}{\beta} \Big/ \sin(\frac{\pi u}{\beta}) \ . \qquad\qquad |u| \geq \varepsilon_F^{-1} \tag{46}$$

When $|u|$ is smaller than ε_F^{-1}, the sum is of the order of $\rho\varepsilon_F$. Using (46) in (45), one obtains G as a function of u_1 and u_2. Note that ΔR also involves u_1 and u_2.

6.3 Path Integral for the Bare Particle

We now discuss the partition function (41). We first neglect S_I. This problem for the bare particle was discussed by COLEMAN [15] in detail, and we will briefly describe his results here.

6.3.1 Stationary Path

We first find a path $R_0(u)$ which makes S stationary and satisfies (33). It satisfies the Euler equation for S:

$$M \frac{d^2 R_0}{du} = \mathrm{grad}U(R_0(u)). \tag{47}$$

One observes that this is the Newton equation for the particle moving in the inverted potential $-U(R)$. With the inverted potential as shown in Fig. 7, one can find several possible solutions of (47). The simplest ones are those in which the particle stays fixed on top of one of the hills:

$$R_0(u) = R_1 \quad \text{or} \quad R_2 \ . \qquad\qquad 0 \leq u \leq \beta \tag{48}$$

6.3.2 Instantons

If the particle is shifted infinitesimally from the top, it will perform a motion as shown in Fig. 8. The small region where R_0 goes from R_1 to R_2 is called an instanton. Its width is about the inverse of the frequency of the

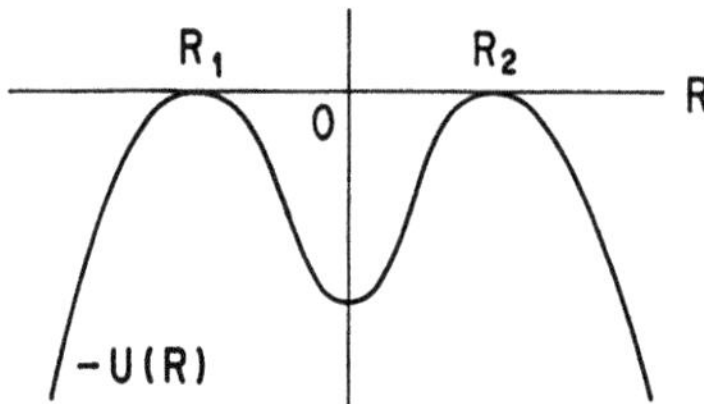

Fig. 7 The inverted potential

12

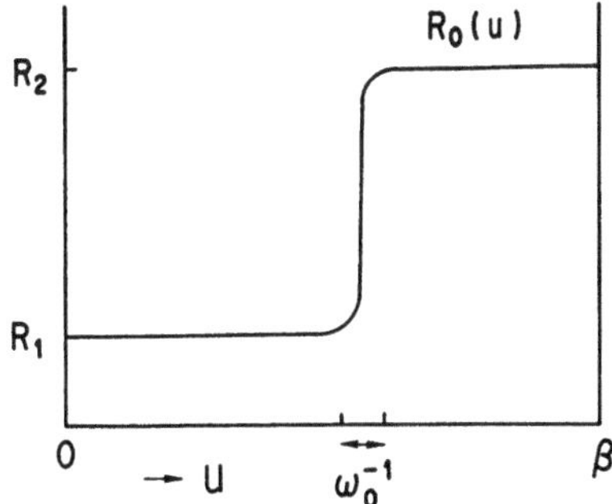

Fig. 8   An instanton

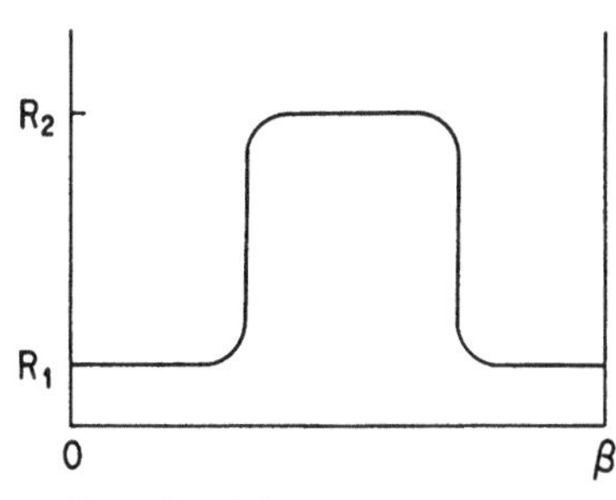

Fig. 9   A bounce

particle in one of the wells (Fig. 4). When an instanton is followed by an anti-instanton, the boundary condition (33) can be satisfied. A pair of an instanton and an anti-instanton is called a bounce (Fig. 9). When β is much larger than the width of the instanton $(1/\omega_0)$, there can be many bounces between 0 and β (Fig. 10). All such paths satisfy the Euler equation and the boundary condition almost exactly.

6.3.3 Path Integrals

Thus, the functional integral reduces to the sum over all such paths as shown in Fig. 10 and the sum over small deviations from such paths. The former consists of the sum over the number of bounces and that over the positions of the instantons and the anti-instantons.

For the constant path, the value of S is zero. [We take $U(\mathbf{R}_1)=U(\mathbf{R}_2)=0$.] The functional integral around this path gives us a factor

$$Z' \propto e^{-\beta\omega_0/2} .$$

For a single instanton, S takes a positive value S_0. The functional integral around this path would give rise to Z', if it were not for a small region around the instanton. The presence of the instanton gives us a correction factor, which we denote by K'. So, the contribution of the instanton is expressed as

$$e^{-S_0} Z'K' = Z'\Delta', \quad \text{where } \Delta' \text{ is defined by}$$

$$\Delta' = e^{-S_0} K' . \tag{49}$$

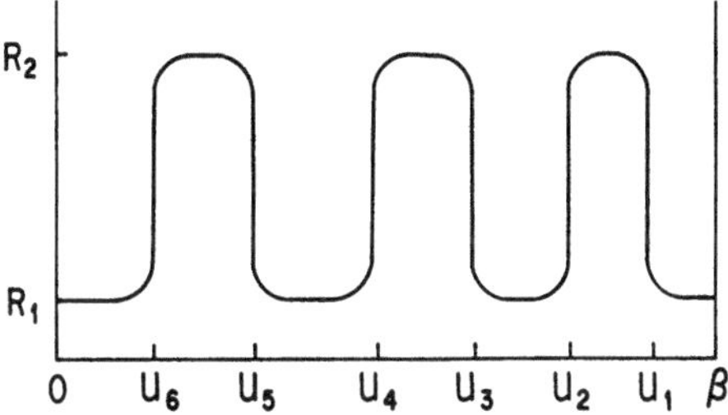

Fig. 10   A possible stationary path

When there are n bounces, each bounce contributes Δ'^2. We must sum over
the positions of the instantons and the anti-instantons. So, we finally find
the partition function of the bare particle as

$$2 \sum_n Z'\Delta'^{2n} \int \cdots \int_{\beta > u_1 > \cdots > u_{2n} > 0} du_1 \cdots du_{2n} = 2Z'\cosh\beta\Delta', \tag{50}$$

where the factor 2 comes from the fact that we may start either from the
left well or from the right well. This is nothing but the partition function
of a two-level system with the separation $2\Delta'$.

6.4 Path Integral for the Interacting Particle

Now, we consider the effect of the electrons. It will be plausible to assume
that in this case, too, the stationary paths consist of bounces, although
the detailed form of the instanton or anti-instanton may be changed. We will
evaluate S_I (43) for a given path.

For the constant path, ΔR is set to zero. In this case, S_I turns out to
be a constant times β, and represents a constant energy shift, which we will
neglect. For a general path of Fig. 10, it is easier to calculate the
difference ΔS_I between this path and the constant path. We easily find

$$\Delta S_I = 2K \sum_{i>j}^{2n} (-1)^{i-j+1} \left(\text{logsin} \frac{\pi(u_j-u_i)}{\beta} - \text{logsin} \frac{\pi\tau_0}{\beta} \right), \tag{51}$$

where K is defined in (20), and u_i is the position of an instanton or an
anti-instanton (Fig. 10).

Furthermore, τ_0 is the cut-off, which is of the order of the width of the
instanton. As I mentioned, the width is about the inverse of ω_0 (Fig. 4).

This is consistent with the recent argument by KAGAN and PROKOF'EV [16].
They argued that the frequency of the particle motion is about ω_0, so the
electronic excitation with frequency higher than ω_0 can follow the particle
motion adiabatically. This fact implies reduction of the non-adiabatic
effect, and ε_F in (31) should be replaced by ω_0. (We naturally assume
$\varepsilon_F \gg \omega_0$.)

I argued [17] that the relevant frequency of the particle should also be
related to the tunneling of the particle, and that the cut-off τ_0 will
be $(N\omega_0)^{-1}$, where N is unity for $k_F a \ll 1$ and $k_F a$ for $k_F a \gg 1$ $(a=|\mathbf{R}_1-\mathbf{R}_2|)$. In any
case, we set the cut-off to 1/D, and use D instead of τ_0.

The action of a given path is the sum of ΔS_I and $2nS_0$. The functional
integral around this path gives us a factor $Z'K''^{2n}$, where K'' may be different from K', which occurred in the case of the bare particle. We define Δ
by

$$\Delta = e^{-S_0} K''. \tag{52}$$

Then, the entire partition function is obtained as

$$Z = Z_{e0} \cdot Z' \cdot 2Z_0 , \quad \text{where}$$

14

$$Z_0 = 1 + \sum_n \Delta^{2n} \int_{\beta > u_1 > \cdots > u_{2n} > 0} \cdots \int e^{-\Delta S_I} \, du_1 \cdots du_{2n}. \tag{53}$$

6.5 An Extension

Our result can be extended to the case where the left and right bottom of
the potential are shifted upwards and downwards by δ. In this case we find

$$Z_0 = \cosh\beta\delta + \sum_n \Delta^{2n} \int \cdots \int \cosh[\delta(\beta - 2(u_1 - u_2 + \cdots))] \cdot e^{-\Delta S_I} du_1 \cdots du_{2n}. \tag{54}$$

From the free energy defined by $F = -kT\log Z_0$, we find the susceptibility χ to
be

$$\chi = -\left. \frac{\partial^2 F}{\partial \delta^2} \right|_{\delta=0} = \frac{1}{\beta} \cdot \langle [\beta - 2(u_1 - u_2 + \cdots)]^2 \rangle. \tag{55}$$

6.6 Comments

6.6.1 Effective Hamiltonian

One might wonder if one can obtain the same result in a much simpler way.
This can be accomplished by introducing an effective Hamiltonian. For this
purpose, we first represent the particle state by a Pauli matrix. Thus,
$\sigma_z = 1$ will represent the particle in the left well, and $\sigma_z = -1$ that in the
right well. Then, σ_x will represent tunneling of the particle between the
wells. The effective Hamiltonian defined by

$$H_{eff} = \Delta \cdot \sigma_x + \delta \cdot \sigma_z + \frac{1}{2}(1 + \sigma_z)H_1 + \frac{1}{2}(1 - \sigma_z)H_2 \tag{56}$$

turns out to give the same partition function as (54), if the cut-off of the
electron energy is set to D instead of ε_F or the bandwidth. Here, H_1 is
defined in (17), and H_2 is similarly defined with $\mathbf{R}_2$ instead of $\mathbf{R}_1$.

6.6.2 Equivalence to the Anisotropic s-d Model

The second comment is that the partition function and the susceptibility of
our system are equivalent to those of another system, namely the anisotropic
s-d model, where the exchange integral for the spin-flip term is different
from that for the spin-non-flip term:

$$H_{sd} = \sum \varepsilon_k c_{k\sigma}^{\dagger} c_{k\sigma} - J_{\parallel} S_z \sum (c_{k'\uparrow}^{\dagger} c_{k\uparrow} - c_{k'\downarrow}^{\dagger} c_{k\downarrow})$$

$$+ 2\mu_B H S_z - J_{\perp} \sum (S_+ c_{k'\downarrow}^{\dagger} c_{k\uparrow} + S_- c_{k'\uparrow}^{\dagger} c_{k\downarrow}). \tag{57}$$

ANDERSON and YUVAL [18] calculated the partition function and the susceptibility of this model by expansion in terms of $J_{\perp}$, and found

$$Z(H=0) \propto 1 + \sum_n J_{\perp}^{2n} \int \cdots \int e^{-S'} du_1 \cdots du_{2n},$$

$$\chi = \frac{\mu_B^2}{\beta} \langle [\beta - 2(u_1 - u_2 + \cdots)]^2 \rangle, \qquad \text{where}$$

$$S' = -(2-\epsilon) \sum_{i>j}^{2n} (-1)^{i-j+1} \left(\text{logsin} \frac{\pi(u_j-u_i)}{\beta} - \text{logsin} \frac{\pi}{\beta\epsilon_F} \right),$$

$$\epsilon = \frac{8\delta_0}{\pi} \left(1 - \frac{\delta_0}{\pi} \right),$$

$$\tan\delta_0 = - \frac{\pi J_\parallel \rho}{2} .$$

So, this model and our system are equivalent, if one assigns parameters of
one system to those of the other as

$$\Delta \rightarrow J_\perp$$

$$2K \rightarrow 2-\epsilon$$

$$\delta \rightarrow \mu_B H$$

$$D \rightarrow \epsilon_F .$$

We want to calculate the partition function (53) in the weak coupling
case (small K), which corresponds to the strong coupling case of the s-d
model ($\epsilon \approx 2$). The exact solution of the s-d model (Sect. 2.5) is applicable
only for the weak coupling case, so we cannot resort to this theory to find
our partition function for small K. In the next section, we will develop a
method to find a solution to this problem.

7. Calculation of the Partition Function

7.1 High-Temperature Expansion

As a preliminary, we take the first two terms of (53). We will see this is a
valid procedure at high temperatures. We find

$$Z_0 = 1 + \Delta^2 \iint \exp\left[-2K\left(\text{logsin} \frac{\pi(u_1-u_2)}{\beta} - \text{logsin} \frac{\pi}{\beta D}\right)\right] du_1 du_2$$

$$= 1 + \Delta^2 \left(\frac{\pi kT}{D} \right)^{2K} \iint_{\beta>u_1>u_2>0} \left(\frac{1}{\sin[\pi(u_1-u_2)/\beta]}\right)^{2K} du_1 du_2 , \qquad (58)$$

where we assumed $D \gg kT$. In the case of the s-d model, such integration as
occurred in (58) is made under the restriction $|u_1 -u_2|<1/D$. This is not
necessary in our case, because K is always less than one half, see (19).
Then, it is possible to change the integration variables to dimensionless
ones and take T out of the integration sign:

$$Z_0 = 1 + \left(\frac{\Delta}{kT} \right)^2 \cdot \left(\frac{\pi kT}{D} \right)^{2K} \cdot \iint_{1>x_1>x_2>0} \left(\frac{1}{\sin\pi(x_1-x_2)}\right)^{2K} dx_1 dx_2. \qquad (59)$$

The integral in the above expression is a constant, which takes the value of
0.5 for K=0 and 0.89 for K=0.3, for example. Then, (59) is equivalent to
high temperature expansion of the partition function of an isolated two-
level system with the tunneling matrix element given by

$$\Delta_{eff} \approx \Delta \cdot \left(\frac{\pi k T}{D} \right)^{K}. \tag{60}$$

The result (59) was first obtained in [11] by a less elegant method. [Note that the susceptibility defined in [11] is different from (55).] We note (60) is the same as (31), except that the cut-off is D instead of the Fermi energy or the bandwidth. The renormalizing factor in (60) represents the effect of the screening electrons on the tunneling motion of the particle.

It is easy to see that the susceptibility obtained in a similar high-temperature expansion is also equivalent to that of an isolated two-level system with the tunneling matrix element given by (60).

7.2 General Case

In order to treat the general terms of (53), we also change the integral variables into dimensionless ones and find

$$Z_0 = 1 + \sum_n \left[\frac{\Delta}{kT} \cdot \left(\frac{\pi k T}{D} \right)^{K} \right]^{2n} \int \cdots \int_{1 > x_1 > \cdots > x_{2n} > 0} \exp\left(-2K \sum_{i>j}^{2n} (-1)^{i-j+1} \log \sin \pi (x_j - x_i) \right) dx_1 \cdots .$$

The integral now depends only on K and n but not on T. We define $c_n(K)$ by

$$c_n(K) = \pi^{2nK} \int \cdots \int_{1 > x_1 > \cdots > x_{2n} > 0} \exp\left(-2K \sum_{i>j}^{2n} (-1)^{i-j+1} \log \sin \pi (x_j - x_i) \right) dx_1 dx_2 \cdots . \tag{61}$$

We also define $\tilde{\Delta}$ by

$$\tilde{\Delta} = \Delta \cdot \left(\frac{\Delta}{D} \right)^{\frac{K}{1-K}} \quad \text{and observe} \tag{62}$$

$$\left[\frac{\Delta}{kT} \cdot \left(\frac{kT}{D} \right)^{K} \right]^{2} = \left(\frac{\tilde{\Delta}}{kT} \right)^{2-2K}.$$

Combining these results together, we obtain

$$Z_0 = 1 + \sum_n c_n(K) \cdot \left(\frac{\tilde{\Delta}}{kT} \right)^{(2-2K)n}. \tag{63}$$

To simplify the expression for the susceptibility, we define $d_n(K)$ by

$$d_n(K) = \pi^{2nK} \int \cdots \int_{1 > x_1 > \cdots > x_{2n} > 0} [1-2(x_1 - x_2 + \cdots)]^{2} \cdot \exp(-2K \sum \cdots) \, dx_1 dx_2 \cdots . \tag{64}$$

Here, the sum in the exponent is the same as that in (61). We find

$$\chi = \frac{\beta}{Z_0} \cdot \left[1 + \sum_n d_n(K) \cdot \left(\frac{\tilde{\Delta}}{kT} \right)^{(2-2K)n} \right]. \tag{65}$$

<u>7.3 Isolated Two-Level System</u>

Equations (63) and (65) are exact results of our model. We first confirm
that they reduce to the known results for the isolated two-level system,
where K=0. Then, c_n and d_n are easily calculated as $c_n=1/(2n)!$ and
$d_n=1/(2n+1)!$. On the other hand, (62) tells us $\tilde{\Delta}=\Delta$. Thus, we obtain

$$Z_0 = \cosh \frac{\Delta}{kT} \,,$$

$$\chi = \frac{1}{\Delta}\cdot\tanh \frac{\Delta}{kT} \,.$$

They are nothing but the partition function and the susceptibility of the
isolated two-level system. (Note that Z_0 is actually half of the partition
function.)

The specific heat C obtained from this partition function becomes
exponentially small at low temperatures, whereas the susceptibility tends to
a constant as the temperatures goes to zero. Then, the Wilson ratio defined
by

$$R = \lim_{T\to 0} \frac{\chi T}{C}\cdot\frac{\pi^2 k^2}{3} \tag{66}$$

is infinite for K=0.

<u>7.4 Independent Bounce Approximation</u>

Here, we will introduce an approximate method to evaluate the coefficients
c_n. This method has been used by GRABERT and WEISS [19] for dynamical problems of the present model. In this approximation, we take only the intra-
bounce interactions between terms in the exponent of (61). Thus, we set

$$c_n(K) = \pi^{2nK} \int\cdot\cdot\int_{1>x_1>\cdot\cdot>x_{2n}>0} e^{-2K\cdot S} \, dx_1 dx_2\cdot\cdot \,, \tag{67}$$

$$S = \sum_{m=1}^{n} \log\sin\pi(x_{2m} - x_{2m-1}) \,.$$

Equation (67) is easily evaluated with the use of the Laplace transform (see
Appendix for a related problem) to be

$$c_n(K) = \Gamma(1-2K)^n/\Gamma(1+2n(1-K)) \,. \qquad (\Gamma: \text{gamma function}) \tag{68}$$

The validity of this approximation will be discussed later for dynamical
problems. Here, we will mention only the results of the approximation and a
refinement of it. Using (68) in (63), we observe that Z_0 is expressed in
terms of the Mittag-Leffler function [20]. The function is defined by

$$E_\beta(z) = \sum_{n=0}^{\infty} \frac{z^n}{\Gamma(1+n\beta)}$$

$$= \frac{1}{2\pi i} \int_{\sigma-i\infty}^{\sigma+i\infty} \frac{e^p}{p}\cdot\frac{1}{1 - (z/p^\beta)}\cdot dp \,. \qquad (\sigma^\beta > |z|) \tag{69}$$

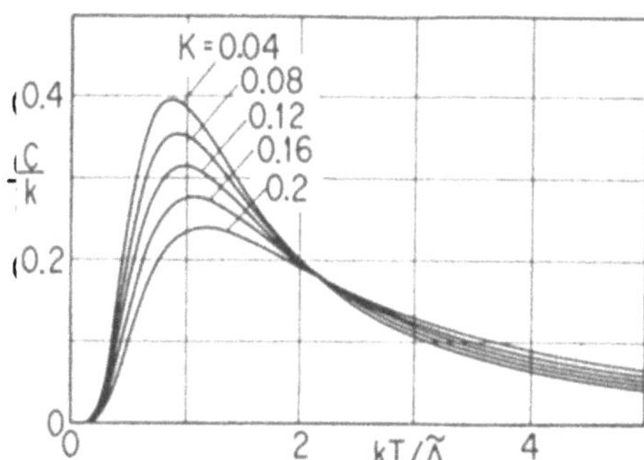

Fig. 11 Specific heats in the independent bounce approximation

So, we see $Z_0 = E_{2-2K}(z)$ with

$$z = \left(\frac{\tilde{\Delta}}{kT} \right)^{2-2K} \cdot \Gamma(1-2K) .$$

The free energy $F_0 = -kT\log Z_0$ and the specific heat $C_0 = -T\partial^2 F_0/\partial T^2$ are easily obtained from Z_0. Figure 11 shows C_0 for several values of K, which shows a Schottky-type anomaly and vanishes exponentially at T=0. Since we know that the specific heat of the s-d model is linear in T at low temperatures, we may naturally expect the same to be true for our model. This means that the independent bounce approximation is not valid at low temperatures.

In order to improve on this point, we assume that the correct c_n is expressed as

$$c_n(K) = \frac{\Gamma(1-2K)^n}{\Gamma(1+2n(1-K))} \cdot \exp\left(g_0 n + g_1 + \frac{g_2}{n} \right) . \tag{70}$$

Our strategy is to calculate c_n numerically for small n (n=1 to 5), and fit the results to (70) to determine g's as a function of K. (The accuracy of numerical calculation of c_n was better than 0.1% for K=0.1 and better than 0.3% for K=0.3.) Figure 12 shows the result of such a fit. We also calculated d_n and fit the results to

$$\frac{d_n}{c_n} = \frac{\gamma_1}{n} + \frac{\gamma_2}{n^2} + \frac{\gamma_3}{n^3} \tag{71}$$

and determined γ's (Fig. 13). (See [21] for the method used here.)

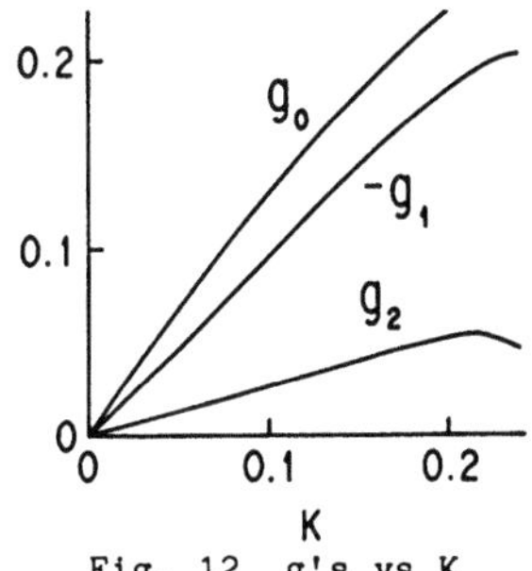

Fig. 12   g's vs K

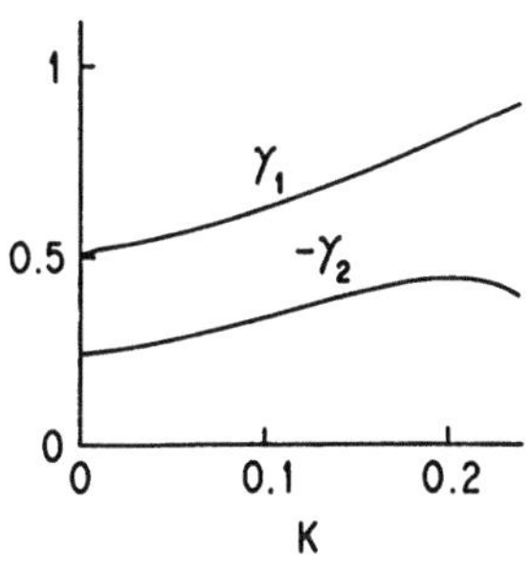

Fig. 13   γ's vs K

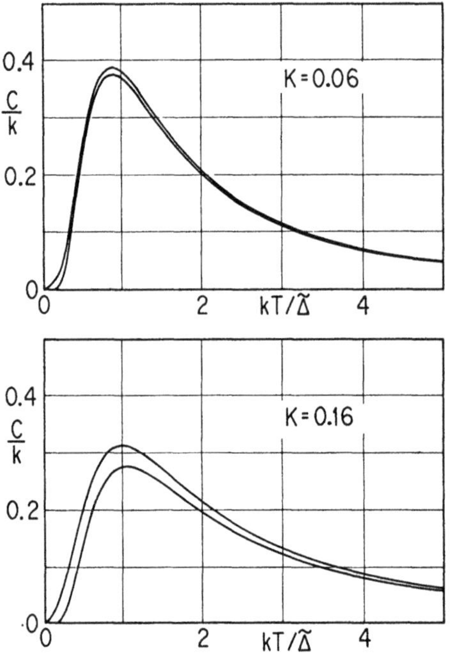

Fig. 14 Specific heat with the use of (70). The lower curves are due to the independent bounce approximation.

7.5 Specific Heat Calculated with (70)

Now, our partition function is obtained from (63) and (70). We can calculate it as well as the specific heat at arbitrary temperatures by taking as many terms as necessary. (We need more terms at lower temperatures.) The results of the specific heat are shown in Fig. 14 for two values of K. We see that our procedure gives an overall feature similar to that of the independent bounce approximation but also gives the specific heat linear in T at low temperatures.

The temperature dependence of the susceptibility is found from (65) and (71) by taking as many terms as necessary both in the numerator and in the denominator of (65). The results are shown in Fig. 15.

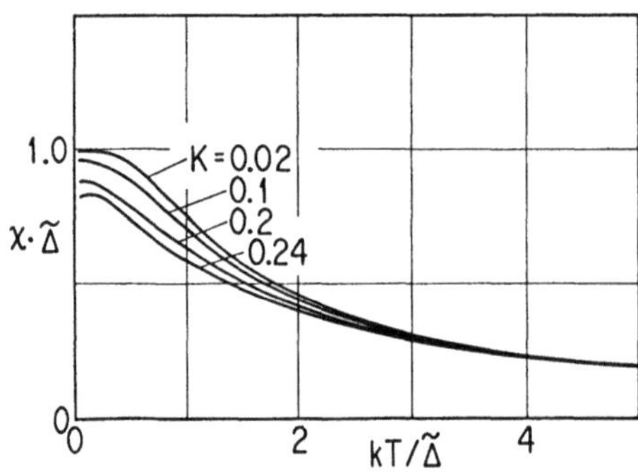

Fig. 15 Susceptibility vs temperature

7.6 Specific Heat at Low Temperatures

We want to analyze the low temperature behavior and find the coefficient of the linear term. (The method used here is found in [21].) We first observe that the nth term of (63) shows a sharp maximum as a function of n, when $\tilde{\Delta}/kT$ is very large. (c_n decreases rapidly with n.) We use Stirling's formula to express the gamma functions in c_n , (70), and find the value of n which gives the maximum:

$$n' = \alpha\tilde{\Delta}/kT, \tag{72}$$

$$\alpha = \frac{1}{2-2K}\cdot\exp\left(\frac{g_0}{2-2K}\right)\cdot\Gamma(1-2K)^{\frac{1}{2-2K}}. \tag{73}$$

We first discuss the susceptibility, which is expressed by the ratio of two sums, see (65). At low temperatures we can take only the largest terms in the sums, i.e., the n'th terms of each sum:

$$\chi = \frac{1}{kT}\cdot\frac{d_{n'}(K)}{c_{n'}(K)} = \frac{1}{kT}\cdot\frac{\gamma_1}{n'} = \frac{\gamma_1}{\alpha}\cdot\frac{1}{\tilde{\Delta}} \ , \qquad (T\to 0) \tag{74}$$

where use is made of (71) and (72). We conclude that the susceptibility tends to a constant of the order of $1/\tilde{\Delta}$ as T goes to zero. This is equivalent to the susceptibility of an isolated two-level system with the effective tunneling matrix

$$\Delta_{eff} \sim \tilde{\Delta} \ . \tag{75}$$

The coefficient γ_1/α is obtained as a function of K from (73) and Figs. 12 and 13. It is unity for K=0 and decreases slowly as K increases.

The specific heat needs more careful treatment. First we determine n' correct up to the first three terms:

$$n' = \frac{\alpha\tilde{\Delta}}{kT} + \frac{p_2}{2-2K} - \frac{1}{2}\cdot\left[\left(\frac{p_2}{2-2K}\right)^2 + \frac{p_5}{1-K}\right]\cdot\frac{kT}{\alpha\tilde{\Delta}} \ . \tag{76}$$

These are the first three terms of the expansion in $kT/\tilde{\Delta}$. We then replace the sum in Z_0 by an integral over n around n'. Assuming that the integral is a Gaussian, we find

$$\log Z_0 = (2-2K)\cdot\frac{\alpha\tilde{\Delta}}{kT} + g_2\cdot\frac{kT}{\alpha\tilde{\Delta}} \ ,$$

from which we obtain the specific heat as

$$C = 2g_2\cdot\frac{k^2 T}{\alpha\tilde{\Delta}}$$

$$\approx 0.92K\cdot\frac{k^2 T}{\tilde{\Delta}} \ . \qquad \text{(small K)} \tag{77}$$

This is linear in T as expected. It is also linear in g_2, which is zero in the independent bounce approximation, see (68). From (77) and (74) we find the Wilson ratio:

$$R = \frac{\pi^2 \gamma_1}{6 g_2} \approx \frac{3.6}{K} . \tag{78}$$

7.7 Interpretation of the Specific Heat Result

Although the low-temperature susceptibility was interpreted as that of an isolated two-level system with the effective tunneling matrix $\tilde{\Delta}$, the low-temperature specific heat cannot be interpreted in the same way. Rather, it represents a contribution from the electrons. Let us examine this point by a simple perturbation calculation. We take the Hamiltonian (56) with $\delta=0$, and regard the potential V as a perturbation. The unperturbed particle state will be represented by $\sigma_x=-1$, if Δ is positive. The σ_z-term of the potential will cause a "spin-flip" scattering

$$k \ \sigma_x=-1 \ \rightarrow \ k' \ \sigma_x=1 ,$$

which will give rise to a self-energy correction for the electron k:

$$E_k = \epsilon_k + \sum_{k'} \left(\frac{V_0}{2}\right)^2 \left| e^{i(\mathbf{k}-\mathbf{k'})\cdot R_1} - e^{i(\mathbf{k}-\mathbf{k'})\cdot R_2} \right|^2$$
$$\times \left(\frac{1 - f_{k'}}{\epsilon_k - \epsilon_{k'} - 2\Delta} - \frac{f_{k'}}{\epsilon_{k'} - \epsilon_k - 2\Delta} \right) .$$

The second term in the parentheses comes from the fact that the process

$$k' \ \sigma_x=-1 \ \rightarrow \ k \ \sigma_x=1$$

is prohibited, when the state k is occupied by an electron. Differentiating this with respect to ϵ_k , setting $k=k_F$ and averaging over the direction of $\mathbf{k}$, we have

$$\left. \frac{dE_k}{d\epsilon_k} \right|_{|\mathbf{k}|=k_F} = 1 - \frac{V_0^2}{2} \cdot \left(1 - \frac{\sin^2 k_F a}{k_F^2 a^2}\right) \cdot \sum_{k'} \left(\frac{1 - f_{k'}}{(\epsilon_{k'} + 2\Delta)^2} + \frac{f_{k'}}{(\epsilon_{k'} - 2\Delta)^2} \right)$$
$$= 1 - \frac{V_0^2 \rho}{2} \cdot \left(1 - \frac{\sin^2 k_F a}{k_F^2 a^2}\right) \cdot \frac{1}{\Delta}$$
$$= 1 - \frac{K}{4\rho\Delta} .$$

This implies a change of the density of states for the electrons

$$\Delta\rho = K/(4\Delta), \quad \text{which causes an excess specific heat}$$

$$C = \frac{\pi^2}{6} \cdot K \cdot \frac{k^2 T}{\Delta} . \tag{79}$$

Comparing (79) with (77), we may interpret (77) as a contribution from the electrons. In (77), the denominator is Δ instead of $\tilde{\Delta}$. This may be due to the effect of higher-order corrections. On the other hand, the correct numerical factor may be $\pi^2/6$ instead of 0.92. This may be due to numerical error in determining g_2. If one uses $\pi^2/6$ instead of 0.92 in (77), the Wilson ratio becomes

$$R = 2\gamma_1/\alpha K \approx 2/K. \tag{80}$$

8. Dynamical Properties of the Two-Level System

We are now concerned with dynamical properties of the two-level system. Suppose that the particle has been fixed in the left well until t=0. Then, we allow it to tunnel to the right well. We are interested in the probability $P_{12}(t)$ that the particle is found in the right well at a later time t, or $P_{11}(t)$ that it remains in the left well. If the system were isolated with the tunneling matrix element Δ_{eff}, P_{12} would be expressed as

$$P_{12}(t) = \sin^2\Delta_{eff}t. \tag{81}$$

The question is how good this is for the system interacting with the electrons, if one uses Δ_{eff} as determined from thermodynamic argument, i.e., (60) at high temperatures and (75) at low temperatures. We will see that (81) is even qualitatively incorrect at high temperatures, whereas it is nearly correct at low temperatures.

8.1 Lifetime of the Particle State

The point is that there is another parameter for dynamical problems, that is the lifetime of the particle state, τ_c. We distinguish two cases. When τ_c is much larger than the period of the oscillation of (81), i.e., $1/\Delta_{eff}$, P_{12} will make many oscillations before it tends to the limiting value 1/2 (Fig. 16), so that (81) will represent the correct behavior at least for short times.

On the other hand, when the reverse is the case, P_{12} will tend to 1/2 before it makes any oscillation (overdamped oscillation: Fig. 17). In this case, P_{12} will be proportional to t for short times:

$$P_{12}(t) = Wt, \tag{82}$$

where the coefficient W is called the hopping rate.

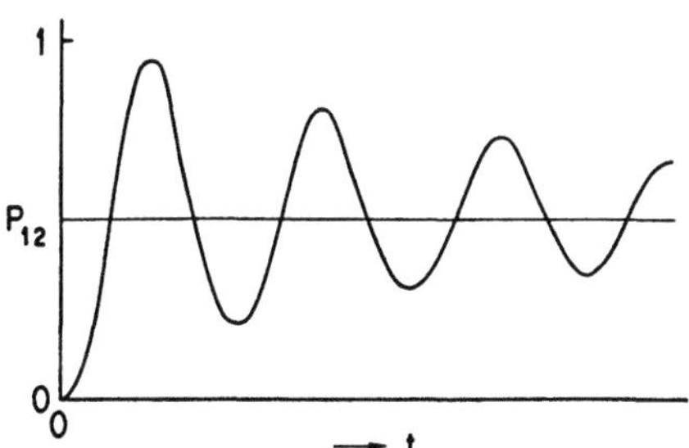

Fig. 16 Underdamped oscillation

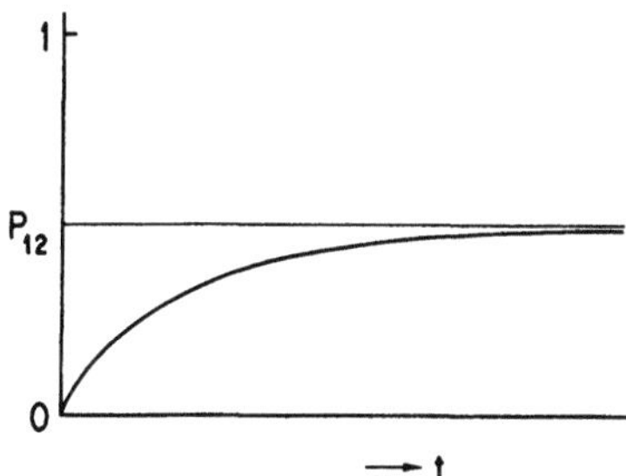

Fig. 17 Overdamped oscillation

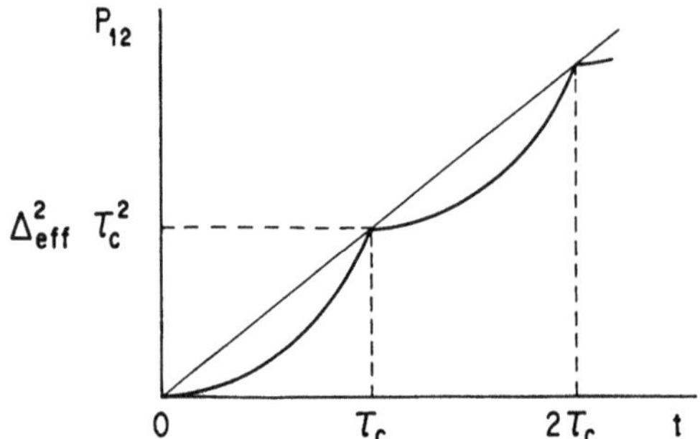

Fig. 18 Effect of damping on hopping rate

The coefficient W can be expressed in terms of τ_c and Δ_{eff}. As is seen in Fig. 18, P_{12} first increases quadratically according to (81), until the particle state is perturbed at a time $\sim\tau_c$. From this time on, P_{12} will again increase quadratically for about τ_c. This process will be repeated and the net result is a linear increase of P_{12} with the coefficient given by

$$W = \Delta_{eff}^2 \tau_c \ . \tag{83}$$

8.2 Perturbation Calculation of τ_c

One may ask how large τ_c is. In order to find the value of τ_c, we will make a perturbation calculation. We again take the Hamiltonian (56) with $\delta=0$, and calculate the transition rate for the particle to jump from the state $\sigma_x=\mp1$ to $\sigma_x=\pm1$ using the golden rule:

$$2\pi \left(\frac{V_0}{2}\right)^2 \sum_{kk'\sigma} \left| e^{i(\mathbf{k}-\mathbf{k}')\cdot\mathbf{R}_1} - e^{i(\mathbf{k}-\mathbf{k}')\cdot\mathbf{R}_2} \right|^2 \delta(\varepsilon_\mathbf{k}-\varepsilon_{\mathbf{k}'}\mp2\Delta)f_\mathbf{k}(1-f_{\mathbf{k}'})$$

$$= \pi K \iint \delta(\varepsilon-\varepsilon'\mp2\Delta)f(\varepsilon)[1-f(\varepsilon')]d\varepsilon d\varepsilon'.$$

This is analogous to the Korringa relaxation of a magnetic spin in metals. We may take this as the inverse of τ_c , so we obtain

$$\tau_c^{-1} = \begin{cases} \pi K kT \ , & kT\gg2\Delta \\ \\ \pi K 2\Delta \ , & kT\ll2\Delta \end{cases} \tag{84}$$

where we considered the case of the lower sign. Later calculation will show that τ_c is actually given by (84), if Δ in that equation is replaced by $\tilde{\Delta}$. Having obtained $1/\tau_c$ and Δ_{eff}, let us see which is larger, so that we can say whether Fig. 16 or Fig. 17 is applicable.

8.2.1 High Temperatures ($kT\gg\tilde{\Delta}$)

In this case, $1/\tau_c$ is given by the upper line of (84), and Δ_{eff} by (60).

24

Then, we have

$$\frac{\tau_c^{-1}}{\Delta_{eff}} = \pi K \cdot \left(\frac{\pi kT}{\tilde{\Delta}}\right)^{1-K}, \tag{85}$$

where use is made of (62). One observes that this is much larger than unity when $kT \gg \tilde{\Delta}$. This corresponds to overdamped oscillation, where W is obtained from (83) as

$$W = \frac{\Delta^2}{\pi K kT} \cdot \left(\frac{\pi kT}{D}\right)^{2K}. \tag{86}$$

8.2.2 Low Temperatures ($kT \ll \tilde{\Delta}$)

In this case, τ_c is given by the second line of (84) with Δ replaced by $\tilde{\Delta}$, and Δ_{eff} by (75). Then, we have

$$\frac{\tau_c^{-1}}{\Delta_{eff}} = 2\pi K. \tag{87}$$

For sufficiently small K, this can be much smaller than unity. In this case, P_{12} will represent an underdamped oscillation.

8.3 Expansion of $P_{12}(t)$ in Δ

We go on to more sophisticated calculation of $P_{12}(t)$. We again take (56) with $\delta=0$ and regard Δ as a perturbation. We assume that initially the particle was in the state $\sigma_z=1$ (which we denote by χ_1) and the electrons in one of the eigenstates of H_1', Φ_1. The wave function at a later time t is expressed by

$$\Psi(t) = e^{-iH_{eff}t} \Phi_1 \cdot \chi_1$$

$$= e^{-iH't}\left(1 + \sum_n (-i\Delta)^n \int \cdots \int_{t>t_1>\cdots>t_n>0} \sigma_x(t_1)\cdots\sigma_x(t_n)dt_1\cdots dt_n\right) \cdot \Phi_1\chi_1, \tag{88}$$

where we expanded the exponential in terms of Δ. Here, $\sigma_x(t)$ is defined by

$$\sigma_x(t) = e^{iH't}\sigma_x e^{-iH't},$$

where H' is defined by

$$H' = \frac{1}{2}(1+\sigma_z)H_1 + \frac{1}{2}(1-\sigma_z)H_2. \tag{89}$$

We note that $H'\chi_1 = H_1\chi_1$ and $H'\chi_2 = H_2\chi_2$, where χ_2 is the state with $\sigma_z=-1$. We also note obvious relations, $\sigma_x\chi_1 = \chi_2$ and $\sigma_x\chi_2 = \chi_1$. Using these relations, we find

$$\Psi(t) = e^{-iH_1t}\left(1 + (-i\Delta)^2 \iint_{t>t_1>t_2>0} e^{iH_1t_1} e^{iH_2(t_2-t_1)} e^{-iH_1t_2} dt_1 dt_2 + \cdots\right) \cdot \Phi_1\chi_1$$

$$+ e^{-iH_2 t}\left((-i\Delta)\cdot\int_0^t e^{iH_2 t_1} e^{-iH_1 t_1} dt_1 + \cdots\right)\cdot\Phi_1\chi_2 . \tag{90}$$

So, the probability that the particle remains in χ_1 irrespective of the electron states is given by

$$P_{11}(t) = \sum_n \langle\Psi(t)|\Phi_n\chi_1\rangle\langle\Phi_n\chi_1|\Psi(t)\rangle$$

$$= 1 - \Delta^2 \int_0^t dt_1 \int_0^t dt_2 \langle\Phi_1|e^{iH_1 t_1} e^{iH_2(t_2-t_1)} e^{-iH_1 t_2}|\Phi_1\rangle + \cdots \tag{91}$$

8.4 Quadratic Term in Δ

First we consider the second term of (91). We assumed that the electrons were initially in the state Φ_1. However, we may assume that they were in thermal equilibrium under the Hamiltonian H_1. In this case the average in the second term of (91) will be replaced by the thermal average:

$$\langle\Phi_1|\cdots|\Phi_1\rangle \to \langle\cdots\rangle_1 = \mathrm{Tr}\{e^{-\beta H_1}\cdots\}/\mathrm{Tr}\{e^{-\beta H_1}\}. \tag{92}$$

Then, the integrand of the second term of (91) is written as

$$Q(t_1-t_2) = \langle e^{iH_1(t_1-t_2)} e^{-iH_2(t_1-t_2)}\rangle_1 . \tag{93}$$

If $Q(t_1-t_2)$ tends to zero sufficiently rapidly as $|t_1-t_2|$ goes to infinity, one may write (91) as

$$P_{11}(t) = 1 - \Delta^2 t\int_{-\infty}^{\infty} Q(t')dt' + \cdots .$$

This means

$$W = \Delta^2 \int_{-\infty}^{\infty} Q(t')dt' . \tag{94}$$

We note $Q(t)$ is a quantity similar to (14), which is written as

$$\langle\Phi_0|e^{-iHt}|\Phi_0\rangle ;$$

$Q(t)$ is a finite temperature version of such quantity. We shall see that it involves all the effects we have been discussing, that is, renormalization of Δ and the lifetime effect.

8.5 Calculation of $Q(t)$ and W [22]

We will evaluate $Q(t)$ by a diagrammatic method. We set

$$H_2 = H_1 + V_{12}. \tag{95}$$

and expand Q in terms of V_{12}, which is the difference of the potential for the electrons between two positions of the particle. In the second quantization scheme, V_{12} is expressed by

$$V_{12} = V_0 \sum_{kk'\sigma} \left(e^{i(\mathbf{k}-\mathbf{k}')\cdot \mathbf{R}_2} - e^{i(\mathbf{k}-\mathbf{k}')\cdot \mathbf{R}_1} \right) c_{k'\sigma}^{\dagger} c_{k\sigma} \, .$$

Then, $Q(t)$ is expanded in terms of V_{12} as

$$Q(t) = 1 + \sum_n (-i)^n \int \cdots \int_{t>t_1>\cdots>t_n>0} \langle V_{12}(t_1)\cdots V_{12}(t_n)\rangle_1 \, dt_1 \cdots dt_n . \qquad (96)$$

The average in the above expression is represented by a set of diagrams. From them we take the most divergent ones, which are represented by such diagrams as shown in Fig. 6. All of such diagrams can be summed to give

$$Q(t) = \exp\left(-2V_0^2 \sum_{kk'} \left| e^{i(\mathbf{k}-\mathbf{k}')\cdot \mathbf{R}_2} - e^{i(\mathbf{k}-\mathbf{k}')\cdot \mathbf{R}_1} \right|^2 f_k(1-f_{k'}) \right.$$

$$\left. \times \frac{1}{(\varepsilon_k - \varepsilon_{k'})^2} [1-\cos(\varepsilon_k-\varepsilon_{k'})t + i\sin(\varepsilon_k-\varepsilon_{k'})t] \right). \qquad (97)$$

We first perform the angular averages and replace the sums by integrals with respect to energies. Keeping the difference of the energies constant, we perform the integral with respect to the sum of the energies and find

$$Q(t) = \exp\left(-2K\int_0^D \frac{1}{\omega}\{(1-\cos\omega t)[1+2n(\omega)] + i\sin\omega t\}d\omega\right), \qquad (98)$$

where ω is the difference of the energies and $n(\omega)$ is the Planck distribution function:

$$n(\omega)=1/(e^{\beta\omega}-1) .$$

As I mentioned previously, the cut-off for the electron energy is set to D. Inserting a factor $\exp(-\omega/D)$ in the integrand and setting the upper limit to infinity, we find

$$Q(t) = \exp\left(-S(t) - 2iK\tan^{-1}Dt\right), \qquad (99)$$

$$S(t) = 2K\log\left(\frac{\sinh\pi kTt}{\pi kTt}\cdot\sqrt{1+(Dt)^2}\right). \qquad (100)$$

We examine limiting cases of $Q(t)$. At zero temperature, we see $Q(t)$ is proportional to t^{-2K} for $t \gg 1/D$. This is analogous to (14). On the other hand, at finite temperatures, $Q(t)$ is approximated by

$$Q(t) = \left(\frac{2\pi kT}{D}\right)^{2K} e^{-2\pi KkT|t|} e^{-2iK\mathrm{sgn}(t)} . \qquad (101)$$

Here, sgn means "the sign of". The first factor is essentially the square of the renormalizing factor in (60), whereas the second factor represents damping with the damping rate $1/\tau_c$. This is what we expected in Sect. 8.2.

Using (99) in (94), we find

$$W = \sqrt{\pi} \, \frac{\Gamma(K)}{\Gamma\left(\frac{1}{2} + K\right)} \cdot \frac{\Delta^2}{D} \cdot \left(\frac{\pi kT}{D}\right)^{2K-1}, \qquad (102)$$

where we assumed kT≪D. This result was obtained by KONDO [22] and YAMADA, SAKURAI and MIYAZIMA [23] and HEDEGARD [14]. The second authors showed that one may use (19) as K in (102). I will make a few comments on (102).

8.5.1 Temperature Dependence

If one expresses W in terms of the golden rule

$$W = 2\pi\Delta_{eff}^2\rho_{eff} \, ,\tag{103}$$

ρ_{eff} will be roughly the inverse of the level broadening. As I mentioned, the latter may be identified with $1/\tau_c$ as calculated in (84). So, ρ_{eff} is inversely proportional to T. Then, (103) tells us that W is proportional to T^{2K-1}. This is what is actually obtained in (102). Since K is always less than one-half, the exponent is negative, so the hopping rate increases as the temperature goes down.

8.5.2 Validity of (102)

Let us examine the condition that our result $P_{11}=1-Wt$ holds. First Wt must be smaller than unity. Second, in deriving (102), we assumed a fast decay of Q(t) and made the following substitution:

$$\int_0^t dt_1 \int_0^t dt_2 \; Q(t_1-t_2) \rightarrow t\int_{-\infty}^\infty Q(t')dt'.$$

As is seen from (101), Q(t') becomes small when $t'\gg\tau_c$. Then, the above replacement is valid when $t\gg\tau_c$. In order that the range of t exists which satisfies both requirements, we must have

$$W \ll \tau_c^{-1} = \pi KkT.\tag{104}$$

Using (62) and (102), one can see that (104) is equivalent to

$$\tilde{\Delta} \ll kT.\tag{105}$$

We have already seen in Sect. 8.2.1 that $P_{12}=Wt$ holds when (105) holds. Furthermore, we note that (102) reduces to (86) for small K.

9. Dynamical Properties of the Two-Level System – Low Temperatures

Now let us consider the general terms of the expansion (91). This part of the talk is mainly due to GRABERT and WEISS [19]. When (90) is used in the first line of (91), we obtain a double summation, the general term of which will look like

$$\int_{t>t_1>\cdots>t_n>0}\cdots\int \int_{t>t_1'>\cdots>t_m'>0}\cdots\int \; < e^{iH_1t_1}\cdots>_1 \; dt_1\cdots dt_n dt_1'\cdots dt_m'.$$

The primed and unprimed variables are ordered, but the order between them is not specified. We express this integral as the sum of integrals, each of which has n+m ordered integration variables. It is more convenient to express the result in terms of P(t):

$$P(t) = 2P_{11}(t) - 1$$

$$= \sum_{n=0}^{\infty} (-2\cos\pi K)^n \Delta^{2n} \int\cdots\int_{t>t_1>\cdots>t_{2n}>0} \exp\left(-\sum_{j=1}^{n} S_j\right)$$

$$\times \sum_{\zeta_1=\pm 1} \sum_{\zeta_2=\pm 1} \cdots \exp\left(\sum_{j>i}^{n} \zeta_j \zeta_i \Lambda_{ji}\right) dt_1\cdots dt_{2n} \qquad (106)$$

where

$$S_j = S(t_{2j-1} - t_{2j}), \qquad\qquad (107)$$

$$\Lambda_{ji} = S(t_{2i}-t_{2j-1}) + S(t_{2i-1}-t_{2j}) - S(t_{2i}-t_{2j}) - S(t_{2i-1}-t_{2j-1}), \qquad (108)$$

and $S(t)$ is defined in (100).

9.1 Dilute (or Independent) Bounce Approximation

The integral in (106) can be performed analytically, if one uses the dilute bounce approximation. In (106) S_j may be considered as the intra-bounce interaction, and Λ_{ji} as the inter-bounce interaction. The approximation amounts to neglecting the latter. This will hold when the bounces are dilute. Due to the factor $\exp(-S_j)$, which will be proportional to $\exp[-2(t_{2j-1}-t_{2j})/\tau_c]$, see (101), the length of each bounce will not be much larger than τ_c. So, when the average length between bounces is larger than τ_c, the positive and negative terms in Λ_{ji} will cancel each other and the dilute bounce approximation will hold. We will later discuss the condition for the approximation to hold. Here, we will evaluate the integral on the basis of the approximation. Details of the calculation will be shown in the Appendix. The result is expressed as

$$P(t) = \frac{1}{2\pi i} \int_{\sigma-i\infty}^{\sigma+i\infty} dp \; e^{pt} \left/ \left[p+2\bar{\Delta}\cdot\left(\frac{\pi kT}{\bar{\Delta}}\right)^{2K-1} \cdot \frac{\Gamma(K+p/2\pi kT)}{\Gamma(1-K+p/2\pi kT)} \right] \right. , \qquad (109)$$

where $\bar{\Delta}$ is the same as $\tilde{\Delta}$ except for a constant factor:

$$\bar{\Delta} = \left(\frac{\sqrt{\pi}\;\Gamma(1-k)}{\Gamma(\tfrac{1}{2}+K)} \right)^{\frac{1}{2-2K}} \tilde{\Delta} \; . \qquad\qquad (110)$$

9.2 High Temperatures

The integral in (109) can be performed in limiting cases. Here, we consider the case of high temperature. Then, $p/2\pi kT$ in the gamma functions may be neglected as compared with K or 1-K. We observe that the quantity

$$\bar{\Delta}\cdot\left(\frac{\pi kT}{\bar{\Delta}}\right)^{2K-1}\cdot\frac{\Gamma(K)}{\Gamma(1-K)}$$

is the same as W defined in (102). Thus, (109) is written as

$$P(t) = \frac{1}{2\pi i} \int \frac{e^{pt}}{p+2W} \, dp = W^{-2Wt}. \qquad\qquad (111)$$

We will see whether our neglect of $p/2\pi kT$ as compared with K is justified. Since (111) tells us $p \sim 2W$, we should have

$$W \ll \pi KkT = \tau_c^{-1},$$

where we used (84). However, as discussed in Sect. 8.5.2, the above inequality is equivalent to (105).

9.3 Validity of the Dilute Bounce Approximation

Let us discuss the validity of the dilute bounce approximation. Equation (111) tells us that the general term of the expansion is $\sim (Wt)^n/n!$. This means that the value of n which makes the most important contribution is $\sim Wt$. Then, this indicates the average distance between the bounces is $\sim t/n \sim 1/W$. So, the dilute bounce approximation will be valid, if this distance is larger than the width of the bounce τ_c.

$$1/W \gg \tau_c.$$

Again this inequality is equivalent to (105).

9.4 Dilute Bounce Approximation at Low Temperatures

According to the above argument the dilute bounce approximation (109) is not valid at low temperatures in general. However, it has recently been argued by LEGGETT, CHAKRAVARTY, DORSEY, FISHER, GANG and ZWEGER [24] that it is valid even at zero temperature unless t is much larger than $1/\tilde{\Delta}$. Let us present another argument to support these authors. We first observe that $S(t)$ in (100) reduces to $2K\log|Dt|$ at zero temperature. Using this result in (106), we have

$$P(t) = \sum_n \left[-2\cos\pi K \cdot \Delta^2 \left(\frac{1}{Dt} \right)^{2K} \right]^n \int_{t>t_1>\cdots>t_{2n}>0} \!\!\!\cdots\int \exp\left(-2K\sum_j \log \frac{t_{2j-1}-t_{2j}}{t}\right)$$

$$\times \sum_{\zeta_1=\pm 1} \cdots \exp\left(\sum_{j>i}\zeta_j\zeta_i\Lambda_{ji}\right)dt_1 dt_2\cdots.$$

On the other hand, the partition function of (53) can be written as

$$Z_0 = \sum_n \left[\Delta^2 \left(\frac{\pi}{D\beta} \right)^{2K} \right]^n \int_{\beta>u_1>\cdots>u_{2n}>0}\!\!\!\cdots\int \exp\left(-2K\sum_j \log\sin\frac{\pi(u_{2j-1}-u_{2j})}{\beta}\right)$$

$$\times \exp\left(\sum_{j>i}\Lambda'_{ji}\right)du_1 du_2\cdots ,$$

where Λ'_{ji} represents the inter-bounce interaction. Comparing these two expressions, one notes close similarity, i.e., t of the first one corresponds to β/π of the second. We have seen, however, that the dilute bounce approximation applied to the latter ($\Lambda'_{ji}=0$) gave a good overall representation of the specific heat except at the lowest temperatures $\beta\gg1/\tilde{\Delta}$. So, we may naturally expect that the dilute bounce approximation applied to the former ($\Lambda_{ji}=0$) will also be a good approximation for $P(t)$ except at large times $t\gg1/\tilde{\Delta}$.

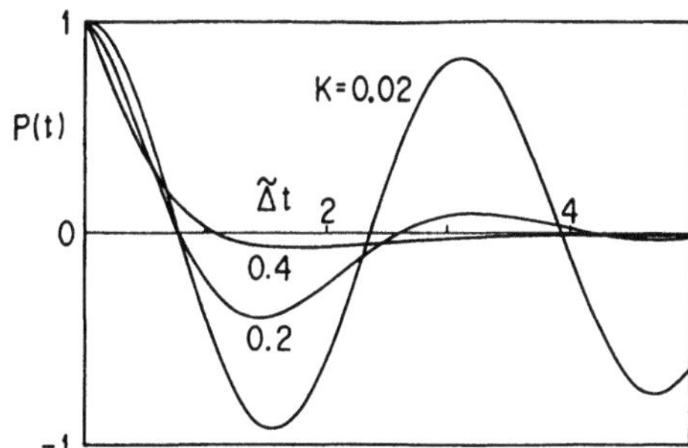

Fig. 19 P(t) for several values of K

9.5 Dilute Bounce Approximation at Zero Temperature

On the basis of the above argument, let us calculate (109) at zero temperature. We first observe

$$\frac{\Gamma\left(K + \frac{p}{2\pi kT}\right)}{\Gamma\left(1 - K + \frac{p}{2\pi kT}\right)} \rightarrow \left(\frac{2\pi kT}{p}\right)^{1-2K} \qquad \text{as } kT \rightarrow 0.$$

Using this result in (109), we find P(t) is written as

$$P(t) = \frac{1}{2\pi i} \int_{\sigma - i\infty}^{\sigma + i\infty} \frac{e^p}{p} \cdot \frac{1}{1 + \left(\frac{2\bar{\Delta}t}{p}\right)^{2-2K}} dp$$

$$= E_{2-2K}(-(2\bar{\Delta}t)^{2-2K}), \tag{112}$$

see (69). Thus, P(t) is expressed in terms of the Mittag-Leffler function of a negative argument. A few examples of P(t) are shown in Fig. 19.

10. Muon Diffusion in Copper

We have discussed dynamic and thermodynamic properties of the two-level system in metals both at high and at low temperatures. We now discuss some experiments which are related to our theory.

This section is about diffusion of the positive muon in copper. Muons are produced with almost 100% spin polarization. The muons injected into copper feel dipolar fields from copper nuclei and their spins precess around the field. Since the directions of the fields at each site are random, the polarization decreases in time. When the muons jump between sites, the dipolar fields are averaged out and the polarization decays with a slower rate. Actually, one measures the decay curve and fits it to a theoretical curve to obtain the jump rate of the muon [25]. Such experiments on the positive muon in copper have been carried out by several groups [26]. An important result of the experiments is that the hopping rate first decreases as the temperature is decreased and then goes up again as the temperature is decreased below ~ 50K.

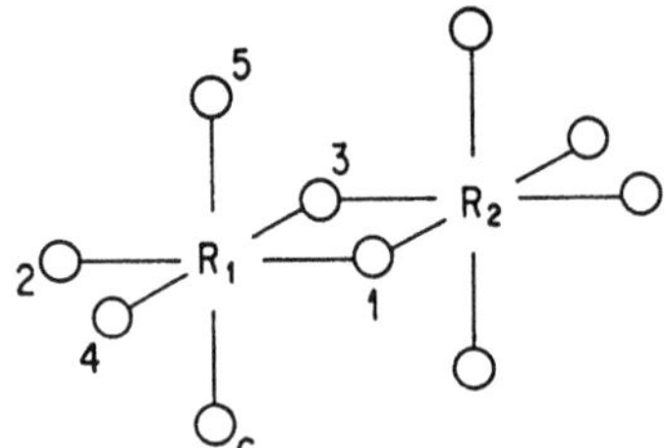

Fig. 20 Muon sites in a fcc lattice

10.1 Theory of the Hopping Rate

We will apply our result (102) for the hopping rate to this system. In order
to explain the experiment, however, one must take account of the interaction
of the muon with lattice vibrations as well as that with the electrons. The
theory of hopping rate of a particle interacting with acoustic phonons has
been developed by many people and is referred to as the small polaron theory
[27]. This theory gives us a hopping rate which decreases as the temperature
decreases and has explained the experiment on the muon in copper in the high
temperature region. It turns out, however, that one must resort to the
theory we have developed so far to explain results in the low temperature
region. We will here develop a theory which takes both interactions into
account.

Suppose that the muon sits in an interstitial site R_1 in Fig. 20. It will
repel the surrounding atoms 1 to 6. Such an effect will be represented by an
interaction

$$H_{L1} = -\lambda(u_{1x} - u_{2x} + u_{3y} - u_{4y} + u_{5z} - u_{6z}),$$ (113)

where u_{1x}, for example, is the x-component of the displacement of atom 1.
Now, we take our Hamiltonian as

$$H = \Delta \cdot \sigma_x + \frac{1}{2}(1+\sigma_z)H_1 + \frac{1}{2}(1-\sigma_z)H_2 , \quad \text{where}$$ (114)

$$H_1 = H_0 + V_{R_1} + \sum_q \omega_q b_q^\dagger b_q + H_{L1}$$ (115)

and H_2 is defined similarly. Here the third term represents the acoustic
phonon energy. The last term can be expressed in terms of the phonon
oprators.

As in Sect. 8, the hopping rate W is expressed by

$$W = \Delta^2 \int_{-\infty}^{\infty} \langle e^{iH_1 t} e^{-iH_2 t} \rangle_1 \, dt,$$ (116)

where the average is now taken over the phonon coordinates as well. Since we
are assuming that the electrons and the phonons are independent, the average
splits into two factors. The phonon part has been treated by the authors of
[27] and the electron part in this lecture, see (99). One just multiplies
both results and integrates over t. In general, the integration must be made
numerically. On the other hand, at low temperatures ($kT \ll \omega_D$), where ω_D is
the cut-off energy for the phonon, we obtain the same result as (102)
with Δ replaced by Δ_0 , which is defined by

$$\Delta_0 = \Delta \, e^{-S}.$$ (117)

Here S is the parameter which occurs in the small polaron theory and represents the strength of the particle-phonon coupling ($S \propto \lambda^2$). Equation (117) implies that the tunneling matrix has been reduced by the small polaron effect. Still, the hopping rate obeys the power law at low temperatures ($\propto T^{2K-1}$). As the temperature is increased, the small polaron effect predominates and the hopping rate increases rapidly.

10.2 Comparison with Experiments

In Fig. 21 we show an experimental result for the hopping rate of the positive muon in copper obtained by BREWER et al. [28]. It shows the tendency mentioned above. We integrated (116) numerically for a choice of parameters. The result is also shown in Figure 21 as a curve. The values of parameters are K=0.18, D=3600K, ω_D=360K, Δ=0.745K and S=4.5. Then, Δ_0 defined in (117) has the value 0.0083K. An important parameter of our theory is $\tilde{\Delta}$, which is defined in (62). One should use Δ_0 instead of Δ in this expression when one takes the small polaron effect into account. Then one finds $\tilde{\Delta} \sim 10^{-3}$K. This is far below the temperatures where the experiment was done, so, our treatment of muon diffusion as a hopping process is justified.

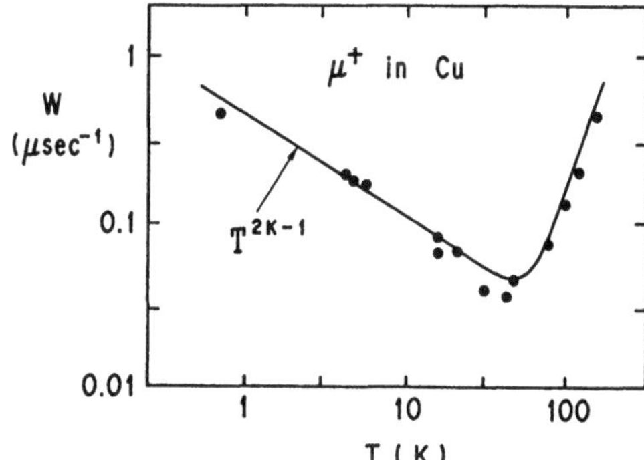

Fig. 21 The hopping rate of the positive muon in in copper [28]

The same system had been investigated by KADONO et al. [29], who had found smaller values of the hopping rate around 50K. Otherwise both results coincide. Their result is shown in Fig. 22. At present, the experimental situation has not been settled, so, we will try to explain the dip found by them. We introduce an interaction which is quadratic in the phonon coordinates. Thus, we add the following term to (115):

$$\frac{1}{2} \mu \left[(u_{1x} - u_{2x})^2 + (u_{3y} - u_{4y})^2 + (u_{5z} - u_{6z})^2 \right].$$
(118)

A similar term is also added to H_2. This interaction, which was first considered by KAGAN and KLINGER [30], causes phonon scattering and so gives rise to a broadening of the muon level, which may be calculated in a similar way to Sect. 8.2. They found that the broadening is proportional to T^9. Since the hopping rate is inversely proportional to the level broadening, we may expect its rapid decrease. Recently, HEDEGARD [31] showed that there exists a term which is expressed as

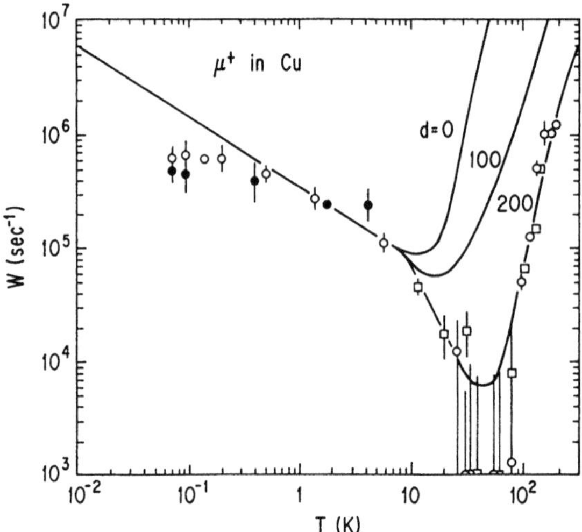

Fig. 22 The hopping rate of the positive muon in copper [29]

$$\frac{1}{2}\,\mu'(u_{1x}^{2} + u_{2x}^{2} + u_{3y}^{2} + u_{4y}^{2} + u_{5z}^{2} + u_{6z}^{2}) \qquad (119)$$

which causes a broadening proportional to T^5. In any case, we may expect a rapid decrease of the hopping rate as the temperature increases.

We add (118) and the corresponding term to H_1 and H_2, respectively and calculate the average in (116) [32]. We first expand the exponential in terms of the interaction in question and take the exponent of the result. This procedure is not fully justified for quadratic interactions such as (118) or (119), but we expect it to be qualitatively correct. Figure 22 contains the result of such a calculation with parameters K=0.2, D=2200K, ω_D=110K, Δ=4K and S=6. The parameter d in the figure is proportional to μ^2 and depends strongly on the Debye cut-off [32]. For some value of d, we obtain a good fit to the experiment. In this analysis, the value of Δ_0 turns out to be 0.01K, which is close to the value in the previous case. The cross-over temperature $\tilde{\Delta}$ is also similar, being roughly 5×10^{-4}K.

11. Proton Tunneling in the Nb(OH)$_x$ System

Since the proton is the lightest atom, one may expect to observe its quantum tunneling in a favorable case, especially at low temperatures. However, when a metal containing hydrogen atoms is cooled down to low temperatures, the hydrogen atoms usually precipitate. It has recently been found that, when niobium metal contains a small amount of oxygen atoms in it, the hydrogen atoms are trapped around the oxygen atoms and remain there down to the lowest tempertaures. They are considered to constitute an ideal two-level system.

A Schottky-type anomaly of the specific heat of this two-level system was observed by WIPF and NEUMAIER [33]. The results were analyzed in terms of the model of a two-level system with the tunneling matrix Δ about 0.2meV and a distribution of δ. The experiment on ultrasonic absorption of this system was also consistent with this model.

34

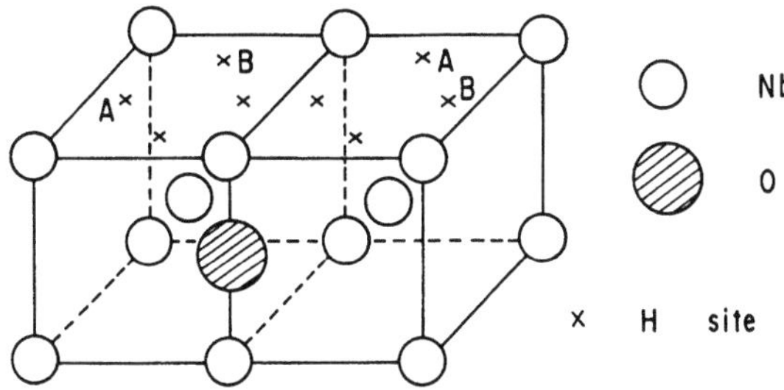

Fig. 23   Interstitial sites in niobium

A crystallographic model was proposed for this system by MAGERL et al. (Fig. 23) [34]. The oxygen atom in niobium sits in the octahedral site as shown in the figure. The hydrogen atoms sit in the tetrahedral sites, which are marked by the crosses. Among them, the pair marked by A and B is considered to be a stable position for the proton, when the oxygen atom occupies the position as marked in the figure. There are many equivalent pairs around an oxygen atom, each of which may constitute a two-level system when it is occupied by a proton.

11.1 Inelastic Neutron Scattering

An experiment on inelastic neutron scattering where the proton is excited from the lower to the upper level of this two-level system was carried out by MAGERL et al. [35]. The structure of the inelastic peak depended strongly on the temperature and it is believed that the interaction with the electrons plays an important part here.

We will make an analysis of neutron data on the basis of the theory developed so far. In general, the differential scattering cross section of neutron is expressed by

$$\frac{d^2\sigma}{d\Omega d\omega} \propto \int_{-\infty}^{\infty} \langle e^{i\boldsymbol{\kappa}\cdot\mathbf{R}(t)} e^{-i\boldsymbol{\kappa}\cdot\mathbf{R}} \rangle e^{-i\omega t} \, dt, \tag{120}$$

where $\boldsymbol{\kappa}$ and ω are the changes of the neutron momentum and the neutron energy, respectively. $\mathbf{R}$ is the proton coordinate. With the use of the Pauli matrix, we obtain

$$\langle e^{i\boldsymbol{\kappa}\cdot\mathbf{R}(t)} e^{-i\boldsymbol{\kappa}\cdot\mathbf{R}} \rangle = \cos^2 \frac{\boldsymbol{\kappa}\cdot\mathbf{a}}{2} + \sin^2 \frac{\boldsymbol{\kappa}\cdot\mathbf{a}}{2} \langle \sigma_z(t)\sigma_z \rangle. \tag{121}$$

The second term of this expression gives rise to an inelastic scattering, and so we are concerned with the quantity defined by

$$I(\omega) = \int_{-\infty}^{\infty} \langle \sigma_z(t)\sigma_z \rangle e^{-i\omega t} \, dt \ . \tag{122}$$

The correlation function which occurs here is related to the hopping rate we have been discussing:

$$P_{12}(t) = \frac{1}{4} \langle [1-\sigma_z(t)](1+\sigma_z) \rangle + \frac{1}{4} \langle (1+\sigma_z)[1-\sigma_z(t)] \rangle. \tag{123}$$

Using (123) and the relation

$$\int_{-\infty}^{\infty} \langle \sigma_z \sigma_z(t) \rangle \, e^{-i\omega t} \, dt = e^{\beta\omega} \int_{-\infty}^{\infty} \langle \sigma_z(t) \sigma_z \rangle \, e^{-i\omega t} \, dt, \quad \text{we find}$$

$$I(\omega) = \frac{2}{1 + e^{\beta\omega}} \cdot 2 \int_{0}^{\infty} P(t) \cos\omega t \, dt, \tag{124}$$

where $P(t)$ is defined in (106).

11.2 High Temperatures

GRABERT et al. [36] developed a theory of neutron scattering from the two-level system on the basis of the theory of Sect. 9.1. To apply (124) to the present case, we resort to our previous results (111) or (112). We first consider (111). Then, we find a quasi-elastic line with the broadening 2W:

$$I(\omega) = \frac{2}{(1 + e^{\beta\omega})} \cdot \frac{4W}{(4W^2 + \omega^2)} \tag{125}$$

11.3 Zero Temperature

In general, when $P(t)$ is expressed in terms of a Laplace transform as

$$P(t) = \frac{1}{2\pi i} \int_{-\infty}^{\infty} e^{pt} Q(p) \, dp, \quad \text{we have}$$

$$2 \int_{-\infty}^{\infty} P(t) \cos\omega t \, dt = Q(i\omega) + Q(-i\omega).$$

With (112), we find $I(\omega)$ has an inelastic peak:

$$I(\omega) = \begin{cases} \dfrac{2}{\tilde{\Delta}} \left(\dfrac{|\omega|}{2\tilde{\Delta}} \right)^{1-2K} \dfrac{\sin\pi K}{\left[\left(\dfrac{|\omega|}{2\tilde{\Delta}} \right)^{2-2K} - \cos\pi K \right]^2 + \sin^2\pi K} , & \omega < 0 \\[20pt] 0 . & \omega > 0 \end{cases} \tag{126}$$

This is close to a Lorentzian form when K is small. We see that the position of the peak is about $\tilde{\Delta}$ away from the elastic peak. We note that (126) is not correct quantitatively near the center of the peak, because it is based on the dilute bounce approximation, which is not valid for $t \gg 1/\tilde{\Delta}$.

We may expect to observe cross-over from (125) to (126) as the temperature is decreased below $\tilde{\Delta}/k$. We note that the cross-over temperature and the energy of the inelastic peak are of the same order of magnitude [37].

11.4 Comparison with Experiments

One sees a clear inelastic peak at 0.05K in [Ref. 35; Fig. 2], which still exists at 5.0K but almost disappears at 9.8K. One complication here is that

niobium is superconducting below 9.2K. As discussed by KONDO [37], RICHTER
[38] and HEDEGARD [31], the existence of a superconducting gap will suppress
effects of the electrons, renormalization of the tunneling matrix and the
level broadening. Recently, WIPF et al. [39] carried out a neutron scattering experiment on normal niobium by applying a magnetic field of 0.7T. They
observed an inelastic peak at about 0.2meV from the central peak at 0.2K,
which almost disappeared at 4.2K. This means that the cross-over temperature
is between 0.2K and 4.2K, which is consistent with the energy of the
inelastic line, that is, 0.2meV. WIPF et al. [39] made a detailed analysis
of their data on the basis of the theory of GRABERT et al. [36].

The proton in the NbO_x system seems to constitute an ideal two-level
system, from which we may obtain more information on the physics of the two-
level system in metals.

Acknowledgement

I would like to thank Professor H. Wipf for showing me reference [39] before
publication.

Appendix

We are concerned with the integral

$$\int_{t>t_1>\cdots>t_{2n}>0} \cdots \int e^{-S(t_1-t_2)} \cdots e^{-S(t_{2n-1}-t_{2n})} \, dt_1 \cdots dt_{2n}. \tag{A.1}$$

As new integral variables, we introduce

$$\tau_i = t_{2i-1} - t_{2i} \qquad \text{and} \qquad \theta_i = t_{2i-2} - t_{2i-1},$$

so that we have

$$t = \tau_1 + \cdots + \tau_n + \theta_1 + \cdots + \theta_{n+1}.$$

Then, (A.1) is rewritten as

$$\int_0^\infty \cdots \int_0^\infty \delta(t-\theta_1-\cdots-\theta_{n+1}-\tau_1-\cdots-\tau_n) \, e^{-S(\tau_1)-S(\tau_2)-\cdots} \, d\tau_1 \cdots d\theta_{n+1}$$

$$= \frac{1}{2\pi i} \int_{\sigma-i\infty}^{\sigma+i\infty} dp \int_0^\infty \cdots \int_0^\infty e^{p(t-\theta_1-\cdots-\tau_n)} \, e^{-S(\tau_1)-\cdots-S(\tau_n)} \, d\tau_1 \cdots d\theta_{n+1}$$

$$= \frac{1}{2\pi i} \int_{\sigma-i\infty}^{\sigma+i\infty} dp \, e^{pt} \left(\int_0^\infty e^{-p\theta} d\theta \right)^{n+1} \left(\int_0^\infty e^{-p\tau-S(\tau)} d\tau \right)^n.$$

The second integral in the above expression can be done analytically:

$$\int_0^\infty e^{-p\tau-S(\tau)} d\tau = \frac{1}{D} \left(\frac{2\pi kT}{D} \right)^{2K-1} \cdot \Gamma(1-2K) \cdot \Gamma\left(K+\frac{p}{2\pi kT}\right) \Big/ \Gamma\left(1-K+\frac{p}{2\pi kT}\right).$$

One observes that the sum in (106) constitutes a geometrical series, which
can easily be summed. In this way, we obtain P(t) as expressed in (109).

<u>References</u>

1. J. Kondo: Solid State Phys. $\underline{23}$, 183 (1969)
   <u>Magnetism V</u>, ed. by H. Suhl (Academic Press, New York 1973)
   G. Grüner and A. Zawadowski: In <u>Prog. Low Temp. Phys. VIIb</u>, ed by D. F.
   Brewer (North-Holland, New York 1978) p.592
2. P. W. Anderson: Phys. Rev. $\underline{124}$, 41 (1961)
3. K. Yamada: Prog. Theor. Phys. $\underline{54}$, 316 (1975) and references therein
   H. Shiba: Prog. Theor. Phys. $\underline{54}$, 967 (1975)
   A. Yoshimori: Prog. Theor. Phys. $\underline{55}$, 67 (1976)
4. K. G. Wilson: Rev. Mod. Phys. $\underline{47}$, 773 (1975)
5. A. M. Tsvelick and P. G. Wiegmann: Adv. Phys. $\underline{32}$, 453 (1983)
6. N. Andrei, K. Furuya and J. H. Lowenstein: Rev. Mod. Phys. $\underline{55}$, 331
   (1983)
7. G. D. Mahan: Phys. Rev. $\underline{163}$, 612 (1967)
8. P. Nozieres and C. T. de Dominicis: Phys. Rev. $\underline{178}$, 1097 (1969)
9. P. W. Anderson: Phys. Rev. Lett. $\underline{18}$, 1049 (1967)
10. K. Yamada, A. Sakurai and M. Takeshige: Prog. Theor. Phys. $\underline{70}$, 73 (1983)
    and references therein
11. J. Kondo: Physica $\underline{84B}$, 40 (1976)
12. J. Kondo: Physica $\underline{124B}$, 25 (1984)
13. F. Guinea: Phys. Rev. Lett. $\underline{53}$, 1268 (1984)
14. P. Hedegard: Phys. Rev. $\underline{B35}$, 533, 6127 (1987)
15. S. Coleman: In <u>The Whys of Subnuclear Physics</u>, ed. by A. Zichichi
    (Plenum Press, New York 1979) p.805
16. Yu. Kagan and N. V. Prokof'ev: Sov. Phys. - J.E.T.P. $\underline{63}$, 1276 (1986)
17. J. Kondo: J. Phys. Soc. Jpn.: $\underline{56}$, 1638 (1987)
18. P. W. Anderson and G. Yuval: Phys. Rev. Lett. $\underline{23}$, 89 (1969)
19. H. Grabert and U. Weiss: Phys. Rev. Lett. $\underline{54}$, 1605 (1985)
20. A. Erdelyi:<u>Higher Transcendental Functions</u> (McGraw-Hill, New York 1955)
    Vol. 3
21. J. Kondo: Physica $\underline{104B}$, 265 (1981)
22. J. Kondo: Physica $\underline{125B}$, 279 (1984); ibid $\underline{126B}$, 377 (1984)
23. K. Yamada, A. Sakurai and S. Miyazima: Prog. Theor. Phys. $\underline{73}$, 1342
    (1985)
24. A. J. Leggett, S. Chakravarty, A. T. Dorsey, Matthew P. A. Fisher,
    Anupam Garg and W. Zwerger: Rev. Mod. Phys. $\underline{59}$, 1 (1987)
25. See, for example, A. Schenck: <u>Muon Spin Rotation Spectroscopy</u> (Adam
    Hilger, Bristol 1985)
26. O. Hartmann, E. Karlsson, L. O. Norlin, T. O. Niinikoski, K. W. Kehr,
    D. Richter, J.-M. Welter, A. Yaouanc and J. LeHericy: Phys. Rev. Lett.
    $\underline{44}$, 337 (1980)
    J.-M. Welter, D. Richter, R. Hempelmann, O. Hartmann, E. Karlsson, L. O.
    Norlin and T. O. Niinikoski: Hyperfine Interact. $\underline{17-19}$, 117 (1984)
    C. W. Clawson, K. M. Crowe, S. E. Kohn, S. S. Rosenblum, C. Y. Huang,
    J. L. Smith and J. H. Brewer: Physica $\underline{109+110B}$, 2164 (1982)
    R. Kadono, J. Imazato, K. Nishiyama, K. Nagamine, T. Yamazaki, D.
    Richter and J.-M. Welter: Hyperfine Interact. $\underline{17-19}$, 109 (1984);
    Phys. Letters $\underline{109A}$, 61 (1985)
27. J. Yamashita and T. Kurosawa: J. Phys. Chem. Solids $\underline{5}$, 34 (1958)
    T. Holstein: Ann. of Phys. $\underline{8}$, 325, 343 (1959)
    C. P. Flynn and A. M. Stoneham: Phys. Rev. $\underline{B1}$, 3966 (1970)
    H. Teichler: Phys. Lett. $\underline{64A}$, 78 (1977)
    D. Emin, M. I. Baskes and W. D. Wilson: Phys. Rev. Lett. $\underline{42}$, 791 (1979)
28. J. H. Brewer, M. Celio, D. R. Harshman, R. Keitel, S. R. Kreitzman, G.
    M. Luke, D. R. Noakes, R. E. Turner, E. J. Ansaldo, C. W. Clawson, K. M.
    Crowe and C. Y. Huang: Hyperfine Interact. 31, 191 (1986)

29. R. Kadono, T. Matsuzaki, K. Nagamine, T. Yamazaki, D. Richter and J.-M. Welter: Hyperfine Interact. $\underline{31}$, 205 (1986)
30. Yu. Kagan and M. I. Klinger: J. Phys. $\underline{C7}$, 2791 (1974)
31. P. Hedegard: preprint
32. J. Kondo: Hyperfine Interact. $\underline{31}$, 117 (1986)
33. H. Wipf and K. Neumaier: Phys. Rev. Lett. $\underline{52}$, 1308 (1984)
34. A. Magerl, J. J. Rush, J. M. Rowe, D. Richter and H. Wipf: Phys. Rev. $\underline{B27}$, 927 (1983)
35. A. Magerl, A. J. Dianoux, H. Wipf, K. Neumaier and I. S. Anderson: Phys. Rev. Lett. $\underline{56}$, 159 (1986)
36. H. Grabert, S. Linkwitz, S. Dattagupta and U. Weiss: Europhys. Lett. $\underline{2}$, 631 (1986)
37. J. Kondo: Physica $\underline{141B}$, 305 (1986)
38. D. Richter: In Quantum Aspects of Molecular Motions in Solids, ed. by A. Heidemann, A. Magerl, M. Prager, D. Richter, T. Springer, Springer Proc. Phys., Vol.17 (Springer, Berlin, Heidelberg 1987) p. 140
39. H. Wipf, D. Steinbinder, K. Neumaier, P. Gutsmiedl, A. Magerl and A. J. Dianoux: To be published in Europhys. Lett.

Singularities in X-Ray Spectra of Metals

G.D. Mahan

Solid State Division, Oak Ridge National Laboratory,
Oak Ridge, TN 37830, USA, and
Department of Physics, University of Tennessee,
Knoxville, TN 37996-1200, USA

1. Introduction

The field of x-ray edge singularities is now about twenty years old. It
started when I predicted this effect [1], although at the time no one
believed me. Two years later they were first observed by HAENSEL et al.[2]
in one of the first experiments utilizing synchrotron radiation as a light
source. Then the phenomena was believed by nearly everybody, and has since
developed into an important field of x-ray physics.

The x-ray spectroscopies we shall discuss are absorption, emission, and
photoemission. The singularities show up in each of them in a different
manner. In absorption and emission they show up as power law singularities
at the threshold frequencies. In photoemission they show up as an
asymmetry in the line shape. In each case the singularities occur for
transitions to or from electron states in the inner core of the atom.
Figure 1 shows the x-ray absorption edge from the 2p-state of sodium metal
as measured by CALLCOTT et al.[3]

This review will emphasize two themes. The first is that the
theoretical model is largely solved. We first proposed a simple model to
describe this phenomena, which is now called the MND model after MAHAN-
NOZIERES-DeDOMINICIS [1,4-5]. Exact analytical solutions are now available
for this model for the three spectroscopies discussed above. These
analytical models can be evaluated numerically in a simple way. Of course,
real physical systems always have a number of complications which are not
described by the simple mathematical model, so that an exact solution is

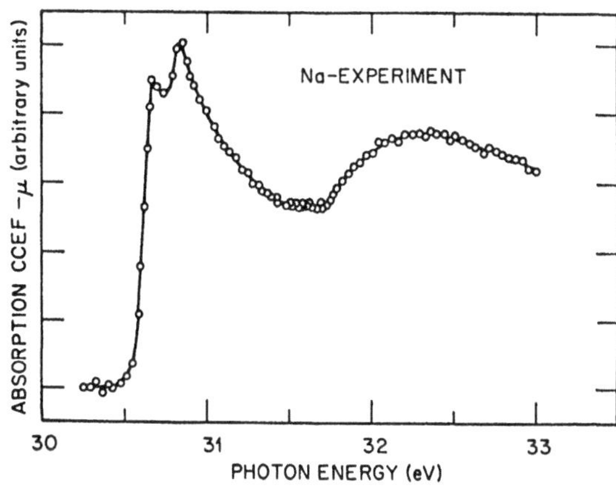

Fig.1 Sodium 2p absorption
spectra

not possible for every system. For example, no model exists which
incorporates correctly the effects of dynamic screening of the electron
gas. This effect is regarded as unimportant for the narrow range of
energies of edge singularities. The most important features of most
physical systems are included in the MND model. So an accurate model can
be solved exactly to describe the singularities.

The second theme of this review is that great care must be used when
comparing the theory to experiment. A number of factors influence the edge
shapes in x-ray spectroscopy. The edge singularities play an important
role, and are observed in many metals. Quantitative fits of the theory to
experiment require the consideration of other factors.

Most of our experimental examples will be limited to simple metals. Edge
singularities may exist in all metals, and are quite important for the
interpretation of x-ray spectroscopies. However, the simple metals have
been the testing ground for the MND theory. They display a wide variety of
phenomena, which provide numerous examples to test the theory.

We take this opportunity to discuss the history of our prediction of
edge singularities. In the middle of the 1960's there were two exciting
many-body phenomena. One was the BCS [6] theory of superconductivity,
while the other was the KONDO effect [7]. COOPER [8] first noted that a
normal metal was unstable to the formation of bound pairs, and his
observation led to the theory of superconductivity. Fig.2 shows a sum of
diagrams which describe the formation of Cooper pairs. Two electrons of
opposite wave vector (k and -k) scatter successively to intermediate and
final states. The vertical dashed lines represent the interaction
potential which causes the scattering. Each pair of intermediate states
causes a vertex correction of the form

$$\Lambda(\omega) = N_F V \int_{-W}^{W} d\xi \; \frac{[1 - 2n(\xi)]}{\xi - \omega} \sim 2N_F V \ln(\omega/W) \qquad (1)$$

where 2W is the band width, N_F the density of states at the fermi surface,
and V the effective interaction potential. There is a logarithmic
dependence upon frequency which occurs because the Fermi occupation factor
$n(\xi)$ is a virtual step function $\theta(-\xi)$ at low temperatures. The summation
of many such vertex terms leads to a pole in the scattering amplitude, as
is described elsewhere in more detail [6].

The Kondo effect has similar mathematics. Here the theory describes how
a conduction electron scatters from a local spin, which is usually provided
by electron states on an impurity. By considering the processes which
include mutual spin flip by the localized and conduction electrons, KONDO
[7] showed that the scattering amplitude had logarithmic terms similar to
those in (1).

One day we realized that logarithmic terms in the scattering amplitude
were still obtained if the two particles in Fig.2 were different. One

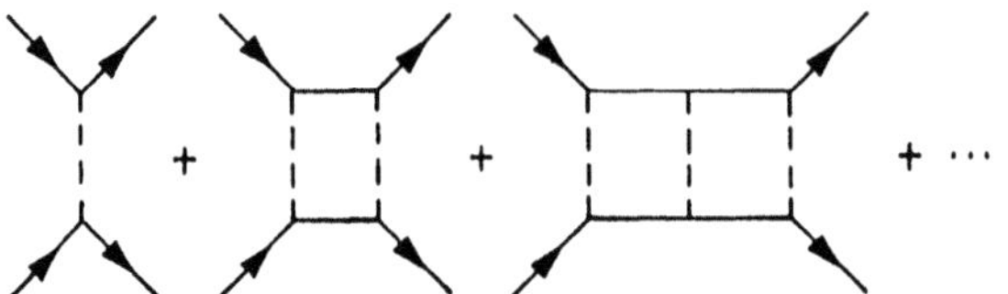

Fig. 2 Feynman diagrams for
pair scattering

could be an electron and the other a hole in a semiconductor. It is important that they have equal and opposite values of wave vector k. In a semiconductor either the electron $\underline{or}$ the hole can be in a filled Fermi sea, but not both. Thus there is only $\underline{one}$ factor of $n(\xi)$ in (1). The factor of two is missing on the right, which is not important. The important idea is that electron-hole scattering in heavily doped semiconductors also has logarithmic singularities which should show up in the interband optical absorption and emission. In those days these experiments were quite popular. There were lots of experimental data, and none at all showed any singularities. I then calculated the effects of quasiparticle damping on the edge singularities, and showed that the singularities were unobservable whenever the particle energy uncertainty $\Gamma=h/(2\pi\tau)$ was larger than 10% of the Fermi energy E_F. That condition is always met in simple semiconductors. One must dope them with impurities to cause the degenerate Fermi sea. The same impurities inescapably cause a short lifetime τ from impurity scattering.

Edge singularities in semiconductors were recently reported in superlattices [9]. Here the impurities are in one layer of the lattice, and the electrons in another, so that the energy width Γ is much reduced.

Our first paper [4] on edge singularities considered them arising from interband transitions in semiconductors. The referee for that paper remarked that the singularities would surely "go away" if the valence band was flat, as it is for a core level. I did this calculation out of curiosity. The singularities became a power law, and did not go away. This calculation became the famous x-ray edge paper [1]. The referee was completely wrong but spectacularly helpful. This calculation was so widely disbelieved at the time that I went through five referees before finding one which would accept it for publication in the $\underline{Physical\ Review}$.

2. MND Theory

We initially proposed a simple model to describe the many-electron phenomena [1,4]. NOZIERES and DeDOMINICIS [5] later treated the same model which is now called MND. There is a system of $N\sim10^{23}$ free electrons which do not mutally interact. Each can interact with a central ion which is undergoing an x-ray transition. In x-ray absorption, an electron in an inner core level is excited to an unoccupied state in the conduction band. Its departure changes the valence of the central ion from Z to Z+1. The N+1 electrons in the conduction band have a screened coulomb interaction with the ion, and react to the change in valence. This reaction can alter the shape of the absorption spectra $A(\omega)$ as a function of the x-ray frequency ω, and cause the excitation of electron-hole pairs in the metal. Here the word 'hole' refers to an excitation of the conduction band with an electron missing from a state beneath the Fermi energy.

An important concept in the MND model is the distinction between the initial and final state Hamiltonians. The initial state Hamiltonian H_i is the summation of two terms: one is for the N noninteracting electrons, and the other is the atomic Hamiltonian describing the bound state of the initial orbital in absorption. The conduction electrons interact with the central potential. This changes the energy of each electron an amount of order $O(1/N)$ which vanishes as N becomes large. The final state potential H_f applies after the x-ray absorption step has created the corehole. It consists of H_i plus the increased potential energy V from the additional charge on the central ion from the corehole. This corehole potential is regarded as screened and therefore short-ranged.

$$H_i = \sum_k \xi_k C_k^+ C_k + \sum_i \xi_i C_i^+ C_i, \quad \xi_k = \frac{k^2}{2m} - \mu, \quad \xi_i = \epsilon_i - \mu,$$

$$H_f = H_i + V, \qquad\qquad V = \frac{1}{N} \sum_{kk'} V_{kk'} C_k^+ C_{k'}. \qquad\qquad (2)$$

The summation over $\underline{i}$ is for bound states. There are also summations over spins which are suppressed.

One electron theory predicts an absorption threshold in metals at an energy equal to the binding energy of the bound state, plus the Fermi energy of the metal, minus a self-energy Δ term from the conduction electrons's interaction with the central ion. Calling this threshold frequency ω_T, MND theory predicts a singular behavior near the absorption threshold of the form

$$A_\ell(\omega) = C_{\ell-1} \left| \frac{W}{\omega - \omega_T} \right|^{\alpha_{\ell-1}} + C_{\ell+1} \left| \frac{W}{\omega - \omega_T} \right|^{\alpha_{\ell+1}} \qquad (3)$$

$$\alpha_\ell = 2\delta_\ell/\pi - \sum_{\ell'} (\delta_{\ell'}/\pi)^2 . \qquad (4)$$

The angular momentum of the initial core state is ℓ, and that of the two possible final conduction band states are $\ell\pm1$. The coefficients $C_{\ell\pm1}$ are the intensities at threshold of the absorption as determined by one-electron theory. The exponents $\alpha_{\ell\pm1}$ determine the power law behavior of the edges. They depend upon the phase shifts δ_ℓ of the conduction electrons of energy E_F for elastic scattering from the screened corehole potential V.

The x-ray singularities occur whenever the exponents α_ℓ are positive. For simple metals this usually happens when $\ell=0$; when the conduction electron is in an s-state. This situation occurs when the initial core electron is in a p-state. For simple metals one generally finds that $\alpha_0 > 0$, $\alpha_1 \sim 0$, and $\alpha_2 < 0$. The x-ray absorption from the outer p-shells of simple metals usually shows singular thresholds. This is the case for all alkali metals except lithium, as well as magnesium and aluminum.

When the initial core state has $\ell=0$, the final conduction state must have $\ell=1$. In this case there are usually negligible edge singularities because $\alpha_1 \sim 0$. They can be neglected in discussing K, L_1, or M_1 spectra.

It is important to remember that there are two terms in eqn.(3). For $\ell=1$ the first term is for s-waves and the second term is for d-waves. Usually the first term is singular at threshold while the second term is rounded. Most experimentalists ignore the second term when fitting their data to the edge singularities. The problem with fitting data is that one needs to know the separate intensities C_0 and C_2 of the edges. These numbers cannot be obtained experimentally, and must be found from theoretical band calculations. A few have been done, and they predict widely different fractions $f=C_2/(C_0+C_2)$ of d-intensity. Figure 3 shows an example where we have averaged an s-edge ($\alpha=0.4$) with a d-edge ($\alpha=-0.2$) to obtain the solid line. The edge singularity is much reduced when the s-and d-edges have equal intensity.

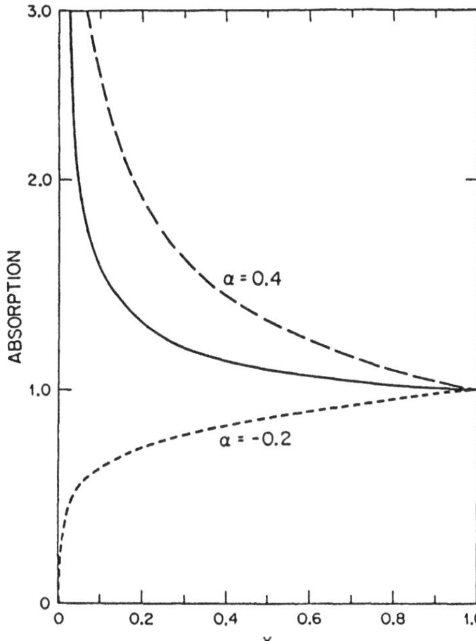

Fig.3 Solid line is average of s-and d-edges in absorption

3. Exact Solution to MND Theory

Several exact solutions have been given to the MND equations. Here we
describe the solution of COMBESCOT and NOZIERES [10] where they reduce the
many-body problem to the evaluation of a matrix. The dimension of the
matrix equals the number of conduction electrons, or $N \sim 10^{23}$. This makes
the matrix hard to evaluate even on a computer. Nevertheless, good
numerical results can be obtained by reducing the number of particles to a
small number such as fifty or one hundred, which then permits numerical
calculation. Here we describe such a calculation, which is done to see the
kind of spectra predicted by MND theory. These calculations produce complete
spectra, and not just asymptotic solutions near threshold.

Interband transitions are introduced to the theory by adding a p·A term
to the Hamiltonian (2). The relevent current operator is

$$j(t) = \sum_{ik} w_i(k)[C_k^+ d_i^+ + d_i C_k] \ .$$

The term $d_i C_k$ destroys an electron-corehole pair, while the other term
creates this pair. The matrix element for optical transitions between
these bands is $w_i(k)$. The time dependent optical absorption A(t) is given
by the current-current correlation function

$$A(t) = \langle|j(t) \, j(0)|\rangle \qquad\qquad (5)$$
$$= \sum_{ijkk'} w_i(k)w_j(k')\langle|e^{iHt}d_i C_k e^{-iHt}C_{k'}^+ d_j^+|\rangle \ .$$

Each current operator has two terms so the product of j·j has four terms.
Only one is interesting for x-ray absorption. We assume that no coreholes

exist in the initial state $|>$, so that $d|>=0$. This eliminates two terms. Our model Hamiltonian has no terms which can scatter a corehole between different states, so the above correlation function must have $i=j$. However, the interaction V in (2) does permit conduction states to scatter between k and k', so that the correlation function A(t) must retain the double summation over k and k'.

The factor exp(iHt) operates to the left on the initial state $<|$ and produces the initial state energy $\exp(iE_g t)$. The other factor exp(-iHt) can be considered to operate to the right. The state $d^+|>$ has one corehole so the conduction band part of H is the final state Hamiltonian H_f. The corehole part of H gives $E_g + \xi_i$. In the MND model (5) is given exactly by

$$A(t) = \sum_{ikk'} w_i(k)w_i(k')e^{-i\xi_i t} F_{kk'}(t) \qquad (6)$$

$$F_{kk'}(t) = <|e^{iH_i t} c_k e^{-iH_f t} c_{k'}^+|> \; . \qquad (7)$$

All of the many-body effects are contained in the correlation function $F_{kk'}(t)$. In the MND model it is assumed to be independent of the corehole i. This approximation is realistic. Experiments indicate that the conduction band interaction with the central atom is independent of which core state contains the hole.

In evaluating this correlation function it is convenient to divide it into two terms

$$F_{kk'}(t) = A_x(t)G(t) \qquad (8)$$

$$G(t) = <|e^{-iH_i t} e^{iH_f t}|> \; ; \qquad (9)$$

G(t) is the corehole Green's function. It describes how the $N \sim 10^{23}$ conduction electrons respond to the appearance of the corehole. ANDERSON [11] first described the renormalization castastrophe which occurs and which is contained in G(t). The sudden appearance of the corehole switches on the final state Hamiltonian H_f. Previous to this time these electrons were in eigenstates of the initial state Hamiltonian H_i. The factor $A_x(t)$ is the exciton part of F. It should have the label kk' but this is dropped to avoid coumbersome notation. The diagrammatic expansion of $A_x(t)$ contains all terms not in G(t). Although this definition seems arbitrary, the division between these two contributions is quite easy in practice.

The exact solution of COMBESCOT and NOZIERES (CN) is expressed in terms of some single particle matrix elements. Let $|k>$ be an eigenstate of H_i with eigenvalue ξ_k. The important matrix element is

$$\phi_{kp}(t) = <k|e^{iH_f t}|p> \; . \qquad (10)$$

This matrix element is difficult to evaluate since $|k>$ and $|p>$ are not eigenstates of H_f. In terms of the exact eigenstates $|\lambda>$ and eigenvalues ξ_λ of H_f one way to evaluate (10) is to insert a complete set of states $1 = \sum_\lambda |\lambda><\lambda|$ and thereby obtain

$$\phi_{kp}(t) = \sum_\lambda e^{i\xi_\lambda t} <k|\lambda><\lambda|p> \; .$$

The corehole Green's function (9) is given by an N-dimensional determinant
of these kinds of matrix elements, where N is the number of occupied
electron orbital states in the conduction band:

$$G(t) = \det\{\phi_{pp'}(t)\} .$$

Each row or column in the determinant is labeled by the wave vector p_i of
an occupied state. The element for the i-th row and j-th column is
$\langle p_i | \exp(itH_f) | p_j \rangle$. This expression is exact but is obviously cumbersome to
evaluate numerically. Similarly the total expression (8) for F(t) is a
similar determinant of dimension N+1. The rows are labeled with the N
occupied states p<F (F is the Fermi wave vector) plus k. The columns are
the N occupied states p<F plus k'. Thus the exact expression for the MND
model is also a determinant of large dimension.

CN showed that the exciton part is

$$A_x(t) = \phi_{kk'}(t) - \sum_{p',p<F} \phi_{kp}(t)[\phi^{-1}(t)]_{pp'}\phi_{p'k'}(t) . \qquad (11)$$

The quantity ϕ^{-1} means the inverse of ϕ, which here is meant to be the
inverse of the matrix of the N×N occupied states. This expression is just
as hard to evaluate as an N×N determinant.

This equation was evaluated by MAHAN [12]. The initial eigenstates were
taken to be unperturbed s-states, so the radial wave function is just the
spherical Bessel function $j_0(kr)$. The corehole potential V(r) was taken to
be a square well, which permits an analytical expression for $V_{kk'}$. A
similar simple function was taken for the optical matrix element w(r).
Thus the only complicated numerical work was in finding the matrix elements
$\phi_{kp}(t)$ and inverting the matrix in (11). Nevertheless the numerical work
is tricky because of the long time behavior. The final expression is the
Fourier transform of (11) which we call $A_x(\omega)$. Taking the transform is
difficult because the convergence is slow at large times. The edge
singularities show divergent behavior at small frequencies, which show up
in $A_x(t)$ as strange behavior at long times. Fig.4 shows typical results.
The numerical noise shows up as emission above the threshold or below the
band edge. Nevertheless the edge singularities appear as a pronounced
feature in the spectrum.

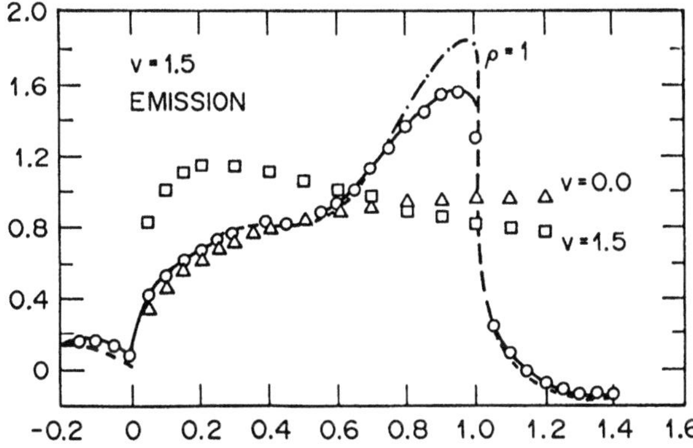

Fig.4 Emission spectra
from solving the CN
equations

There have been many numerical solutions of the MND equations. Most
take a band of finite width which limits the number of states. One early
important calculation was by GREBENNIKOV et al. [13]. A different
approximation was made by VON BARTH and GROSSMANN [14]. They assumed that
the potential was separable $V_{kk'}=g_k g_{k'}$. This has the effect of reducing
the conduction electrons to local operators $c(r)$ on the impurity

$$\sum_{kp} V_{kp}\, c_k^+ c_p = \{\sum_k g_k c_k^+\}\{\sum_p g_p c_p\} = c^+(r)c(r)$$

$$c(r) = \sum_k g_k c_k .$$

This approximation simplifies the numerical work. Mahan's calculation is
the only one so far with an unlimited band width and a nonseparable
potential. Below we show that the separable potential model can be solved
analytically.

One reason for doing these calculations is to test the hypothesis of the
Final State Potential. We first proposed this idea [12]. The idea comes
from ignoring the edge singularities and examining the other parts of the
absorption or emission spectra. Should they be calculated using H_i or H_f?
Before answering this question, first note how these two choices enter into
absorption and emission:

	Initial	Final
Absorption	H_i	H_f
Emission	H_f	H_i

In emission the initial state has the corehole (H_f) while the final state
has no corehole potential (H_i). This is reversed in absorption. A
consistent choice would be to choose the final state in both emission or
absorption, or else the initial state Hamiltonian in both cases. However,
this is not usually done. Most band theorists calculate wave functions for
the ground state of the solid, and use those wave functions for calculating
both absorption and emission. Thus they are using the initial state
Hamiltonian for absorption and the final state Hamiltonian for emission.

For emission calculations one should use the final state Hamiltonian,
which are the wavefunctions calculated without the presence of the
corehole. This choice is demonstrated by two kinds of calculations. The
first are one-electron emission calculations done with and without
coreholes. Those without coreholes produce spectra which resemble
experiments. Adding a screened corehole changes the s-state wavefunctions,
and produce quite different spectra. These unusual spectra look just like
the experiments on satellite bands. This is a further confirmation of the
final state rule, since the final state of satellite spectra contain one
core hole.

The second calculation which suggests the final state rule for emission
is the solution of the MND equations as shown in Fig.4. Away from the edge
singularities, the calculated spectra have the same shape as given by the
one-electron spectra without the coreholes. This one-electron shape is
shown as the triangles. The squares are the one-electron spectra with the
corehole, and would correspond to the initial state rule. The numerical
solution of the MND equations also support the final state rule.

For x-ray absorption the rule is less obvious. Consistency suggests
that one should also use the final state rule. However, calculations with

and without the screened corehole differ little in absorption when done for
free-electron metals. Here the case is less convincing. However, another
set of states was proposed by DAVIS and FELDKAMP [15] which seems
successful [16].

4. Dispersion Theory

Another exact solution of the MND equations is based upon dispersion
theory. The first use of this technique for edge singularities was by
PARDEE and MAHAN [17]. They showed how to derive the edge singularities
using only a few lines of algebra. This simple approach was disbelieved
and ignored for a decade. Then PENN, GIRVIN and MAHAN [18-19] (PGM) solved
the scattering equations exactly. For a separable or constant potential
they found that the transition amplitude was given by dispersion integrals,
in agreement with Pardee and Mahan. For these simple potentials, they used
dispersion integrals to write down exact expressions for both the exciton
part $A_x(t)$ and the corehole Green's function $G(t)$. This was a second exact
solution, following the earlier one by Combescot and Nozieres. Both $A_x(t)$
and $G(t)$ were expressed as infinite series, where each term in the series
was specified, and where the series has rapid convergence. Then OHTAKA and
TANABE [20-23] used a similar technique to derive the exact solution for
the MND equations for any kind of potential. Their solution was in closed
form, rather than a series.

Here we reproduce the simple argument of Pardee and Mahan. Let $T(\omega)$ be
the complex scattering amplitude for an ingoing wave of an electron
scattering from a central potential, which in this case is the screened
corehole. If $T'(\omega)$ is the amplitude of the outgoing scattered wave, then
they differ only by a phase factor which is the phase shift

$$T'(\omega) = T(\omega) \exp[2i\delta(\omega)\Theta(\omega-\omega_T)] .$$

We omit the subscript ℓ on the phase shift, although it does belong since
these arguments apply to each angular momentum channel ℓ. The step
function $\Theta(\omega-\omega_T)$ reminds us that the phase factor only exists in the region
of absorption, since both T and T' must be real in regions where there is
no absorption. The next assumption is that T and T' both originate from
the same analytical function of frequency:

$$t(\omega+i\delta) = T(\omega)$$
$$t(\omega-i\delta) = T'(\omega) .$$

Combining these results, we learn that

$$t(\omega+i\delta) = \exp[\phi(\omega+i\delta)] = |T(\omega)| \exp[i\delta(\omega)\Theta(\omega-\omega_T)]$$

from which we deduce

$$\mathrm{Im}[\phi(\omega\pm i\delta)] = \pm\delta(\omega)\Theta(\omega-\omega_T) . \qquad (12)$$

What kind of function obeys (12)? The answer from dispersion theory is
that the only analytical function of frequency which has this property is

$$\phi(\omega+i\delta) = \frac{1}{\pi} \int_{\omega_T}^{W} d\varepsilon \, \frac{\delta(\varepsilon)}{\varepsilon-\omega-i\delta} .$$

The function $\phi(\omega+i\delta)$ has the form of a dispersion integral. For
frequencies near threshold this can be approximated as

$$\phi(\omega+i\delta) \sim [\delta(\omega_T)/\pi]\ \ln[W/(\omega_T-\omega-i\delta)]$$

$$|t(\omega+i\delta)| \sim \left|\frac{W}{\omega-\omega_T}\right|^{\delta/\pi},$$

where $\delta(\omega_T)$ is the phase shift for conduction electrons at the Fermi surface. The above results only apply near the threshold. However, they show that the dispersion integral has the correct singular behavior at threshold. According to dispersion theory, this singular behavior is just a consequence of having a step in the one-particle absorption spectra. This step is caused by the sharpness of the Fermi surface in energy space. These arguments show that any system with a step in the one-particle spectra will have a power law behavior at that step.

Dispersion theory [24] was developed for scattering theory in nuclear and high energy physics. The scattering amplitude as a function of particle kinetic energy has distinctive variations whenever a new channel appears in the cross section: e.g., when a new particle can be created. The x-ray edge singularities are quite well described by this theory. They are another kind of threshold phenomena, of the kind well known in scattering theory.

The earliest theories of the edge singularity suggested that the phenomena resulted from an infrared divergence due to the creation of multiple electron-hole pairs during the absorption or emission. The dispersion theory of Pardee and Mahan makes no reference to electron-hole pairs, yet derives the singularities. Everyone believed in the infrared divergence, and therefore disbelieved Pardee-Mahan.

This confusion was removed by PGM. They pointed out that only the core-hole Green's function $G(t)$ had an infrared divergence. The exciton intensity $A_x(t)$ does not have an infrared divergence. They showed that the exciton term can be expanded in a series

$$A_x(\omega) = \sum_{n=0}^{\infty} A_n(\omega). \tag{13}$$

The n-th term in this series is the probability of the exciton transition occurring with the creation of n pairs of electron-holes. Each term in this series is finite for all frequencies above threshold--it has no infrared divergence. Furthermore, the series converges rapidly. In typical cases the size of each successive term decreases tenfold from the preceding one. One percent accuracy can be obtained by just retaining A_0+A_1.

PGM showed that the Pardee-Mahan result was just A_0, which is the absorption probability for zero pairs. Here we use the phrase 'zero pairs' or 'one pair' to discuss only the exciton term $A_x(\omega)$. The corehole Green's function $G(\omega)$ has an infrared divergence from the creation of an infinite number of low energy electron hole pairs.

$$A_0(\omega) = w^2 |t(\omega+i\delta)|^2,$$

where w^2 is the matrix element found from one-electron theory. For $n>0$ the intensities correspond to making n pairs. For each electron involved a similar factor is needed for holes. They have their energy below the Fermi surface, so their dispersion integrals span that energy space. We define separate dispersion integrals for electrons t_e and holes t_h

$$t_e(\omega) \equiv t(\omega+i\delta) = \exp[\frac{1}{\pi} \int_{\omega_T}^{W} d\varepsilon \frac{\delta(\varepsilon)}{\varepsilon-\omega-i\delta}]$$

$$t_h(\omega) = \exp[\frac{1}{\pi} \int_{-W}^{\omega_T} d\varepsilon \frac{\delta(\varepsilon)}{\varepsilon-\omega-i\delta}] \ .$$

$$(14)$$

The exact amplitudes A_n can be written down easily. One just writes the expressions from the Golden rule, and then adds the factors of t_e for each electron state and t_h for each hole state. If $\rho(\varepsilon)$ is the one-particle density of states, then expressions for the first two terms are

$$A_0(\omega) = 2\pi w^2 \int_{\omega_T}^{W} d\varepsilon \rho(\varepsilon) |t_e(\varepsilon)|^2 \delta(\varepsilon-\omega)$$

$$A_1(\omega) = 2\pi w^2 \int_{\omega_T}^{W} d\varepsilon_1 \rho(\varepsilon_1) \int_{-W}^{\omega_T} d\varepsilon_2 \rho(\varepsilon_2) \int_{\omega_T}^{W} d\varepsilon_3 \rho(\varepsilon_3) |t_e(\varepsilon_1) t_h(\varepsilon_2) t_e(\varepsilon_3)|^2$$

$$\times \frac{\delta(\varepsilon_1-\varepsilon_2+\varepsilon_3-\omega)}{(\varepsilon_1-\varepsilon_2)(\varepsilon_2-\varepsilon_3)} \ .$$

In A_1 the final state has two electrons above the Fermi surface (ε_1 & ε_3)and one hole below (ε_2). The two electrons are indistinguishable and enter the equation symmetrically. PGM wrote down the general expression A_n in the same fashion. Thus they found an exact expression for $A_x(\omega)$. Figure 5 shows a calculation of the frequency dependence of A_0 and A_0+A_1 taken from ref. 18. The one pair term A_1 makes only a small correction to the exciton intensity.

MAHAN [19] used the same techniques of PGM to evaluate the corehole Green's function. He showed that

$$G(t) = -i\theta(t) \ P(t) \ \exp[-it(\varepsilon_h + \Delta)]$$

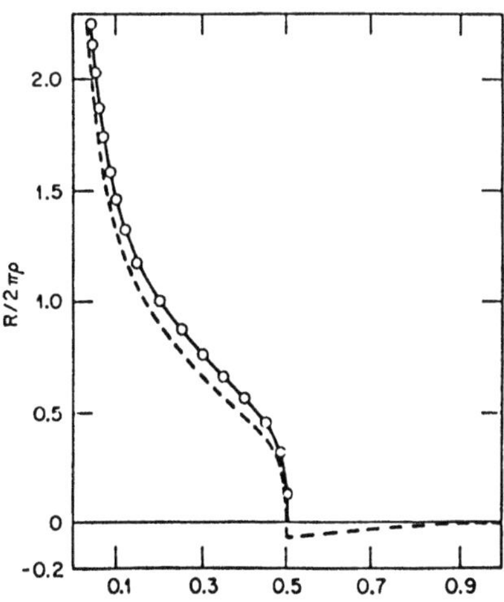

Fig.5 Zero and one pair intensities

$$P(t) = \exp[-\phi(t)] \tag{15}$$

$$\phi(t) = \int_0^\infty \frac{du}{u} R(u) (1 - e^{-itu}) . \tag{16}$$

The quantity $P(t)$ can be expressed as an infinite series of terms, where the n-th term corresponds to the creation of n pairs. Each term has an infrared divergence, and Fourier transforming each individual term leads to a nonsensible mathematical result. Instead, it is necessary to employ a cumulant expansion to resum the divergent terms. One then obtains the exponential factor $\phi(t)$. The cumulant expansion produces an expression for $R(u)$ which is also expressed as a series

$$P(t) = \sum_{n=0}^\infty P_n(t)$$

$$R(u) = \sum_{n=1}^\infty R_n(u) .$$

The final expression for $P(t)$ can be Fourier transformed, although not the individual terms $P_n(t)$. The cumulant expansion permits the summation of the most divergent terms in the series which gives an expression which can be Fourier transformed.

The constant P_0 is the probability of no pairs being created. The one pair contributions to P_1 and R_1 found by Mahan are

$$P_1(t) = P_0 V^2 \int_0^F \frac{dk_1}{\pi} \int_F^\infty \frac{dk_2}{\pi} \frac{|t_e(\epsilon_1)t_h(\epsilon_2)|^2 \, e^{-it(\epsilon_2-\epsilon_1)}}{(\epsilon_1 - \epsilon_2)^2}$$

$$R_1(u) = \frac{V^2}{u} \int_0^F \frac{dk_1}{\pi} \int_F^\infty \frac{dk_2}{\pi} |t_e(\epsilon_1)t_h(\epsilon_2)|^2 \delta(u+\epsilon_1-\epsilon_2) ,$$

where V is the electron-hole interaction, F is the Fermi wave vector, and ϵ_1 and ϵ_2 are the band energies associated with wave vectors k_1 and k_2. Both of these expressions contain dispersion integrals. Dispersion theory plays a central role in the evaluation of both $G(t)$ and $A_x(t)$.

Mahan obtained the exact expression for the n-th term $P_n(t)$ but not for $R_n(u)$.

OHTAKA and TANABE [20-23] carried this approach much further. They obtained analytical expressions for $P(t)$ and $R(u)$ which apply for any form of electron-hole potential--not just for the separable or constant potential of PGM. Their expressions are in the form of determinants of N-dimensional matrices, so are similar in format to the exact solution of CN. They have evaluated their expression and obtained a number of new results. One of the simplest and most important is an expression for the renormalization of the intensity at the singular edge. PGM showed previously that the asymptotic expression near threshold is <u>not</u> given by the MND form (3). The threshold constant C_ℓ is renormalized by a factor of z because of the emission of electron-hole pairs. Each angular momentum channel has the threshold behavior of

$$A(\omega) = C_\ell z_\ell \left| \frac{W}{\omega-\omega_T} \right|^{\alpha_\ell} .$$

PGM obtained an exact but cumbersome expression for z_ℓ. Ohkata and Tanabe showed that it has the simple expression

$$z_\ell = \frac{\Gamma(1-\delta/\pi)^2}{\Gamma(1 - 2\delta/\pi + (\delta/\pi)^2)} \, ,$$

where the phase shifts also depend upon ℓ. This factor arises from the inclusion of multipair final states, which was omitted by all of the earlier workers in the field. The factor z_ℓ is typically only a few percent different than one, but should be included in any careful fits of experiment to theory.

5. Incomplete Phonon Relaxation

This review has mostly discussed the theory of x-ray absorption. The emphasis on absorption occurs because the edge singularities show up best in this experiment. The same phenomena are expected to show up in x-ray emission. However, the interpretation of the emission spectra is complicated by the process of incomplete phonon relaxation.

After the creation of the corehole, the electrons in the metal adjust rapidly to this new potential. This adjustment is called 'screening'. The time scale for the formation of the screening charge is the inverse of the plasma frequency ω_p of the metal. This time scale applies to the contributions of both plasmons and electron-hole pairs. The time scale for the x-ray emission is determined by the corehole lifetime which is mainly limited by Auger scattering. Usually the Auger lifetime τ obeys the relation $\omega_p\tau \gg 1$ and the emission takes place from a fully relaxed electronic state.

The ions are much heavier than the electrons and take much longer to adjust to the presence of the new core hole. Ionic motion or relaxation is governed by the phonon modes of the solid. The typical time for adjustment of the phonon modes is the inverse of the Debye frequency ω_D. The phonon system is fully relaxed around the corehole whenever $\omega_D\tau \gg 1$. This latter condition is rarely satisfied in simple metals. The x-ray emission usually occurs from a state of incomplete phonon relaxation around the screened corehole. Alternately, one can state that the electronic parts of the screening are relaxed, but not the ionic parts. Incomplete Phonon Relaxation (IPR) complicates the interpretation of the x-ray emission spectra. For metallic lithium IPR is the dominant contribution to the deviations of the spectra from one-electron theory. Here IPR is much more important than edge singularities or Auger lifetime effects. For other metals IPR plays a lesser but still important role. For example, edge singularities are usually less sharp in emission than in absorption because of IPR. The influence of IPR makes emission harder to interpret than absorption.

The theory of IPR was first developed by MAHAN [25] and independently by ALMBLADH [26]. The second Born approximation is used to treat absorption and emission as a connected two-step process. The corehole is assigned an Auger lifetime of γ which provides exponential decay in time. The lineshape function $E(\omega)$ for emission is given by the following function if one assumes that the corehole was created by broadband absorption of photons

$$E(\omega) = 2\mathrm{Re}\left\{\int_0^\infty d\tau_1 I(\tau_1) \exp[-i\tau_1(\omega+2\Sigma-i\gamma) - \phi(\tau_1)]\right\}$$

$$I(\tau_1) = \int_0^\infty d\tau_2 \exp\{-2\gamma\tau_2 - 2i\sum_\lambda (M_\lambda/\omega_\lambda)^2 [\sin(\omega_\lambda\tau_2) - \sin\omega_\lambda(\tau_1+\tau_2)]\}$$

$$\phi(\tau) = \sum_\lambda (M_\lambda/\omega_\lambda)^2 \{(N+1)[1-i\omega_\lambda\tau-\exp(-i\omega_\lambda\tau)]+N[1+i\omega_\lambda\tau-\exp(i\omega_\lambda\tau)]\}$$

$$N = 1/[\exp(\beta\omega_\lambda)-1].$$

The line shape function $E(\omega)$ must be convoluted with the one-particle
spectra and other many-body contributions in order to get the final
emission spectra. $E(\omega)$ only contains the effects of phonons. The self-
energy $\sum$ is from phonons and $2\sum$ is the usual Franck-Condon shift. The
summation over λ is over the phonon modes and wave vectors, where ω_λ are
the phonon frequencies and M_λ is the matrix element between the phonons and
the corehole. The summation over λ should be changed to a continuous
integral over the phonon spectra. The upper limit of this integral is the
Debye energy ω_D.

The computed spectra do not depend upon the detailed shape of the phonon
density of states nor upon the frequency dependence of the matrix element.
The two main parameters are the overall coupling strength and the Debye
energy. It is important that the phonon bands have a finite width. A
number of phonons are created during the absorption step. Phonon relaxation
occurs by having these phonons die out (damping), or wander away
(dispersion). Either damping or dispersion are needed for phonon
relaxation. They enter the theory in a similar way and cause similar
effects. Without them the phonons would oscillate forever and never relax.

Figure 6 shows the type of effects which can occur. Here we modeled the
lithium spectra by assuming a Lorentzian shaped phonon density of states
centered at $\omega_0=0.05$ eV and with a width $\Gamma=0.05$ eV. Part (a) shows three
lines. The solid line is the computed absorption spectra, which is a

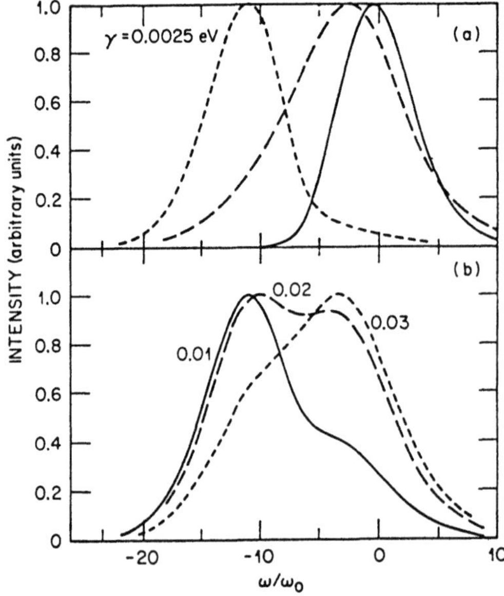

Fig.6 Emission lineshape
for incomplete relaxation

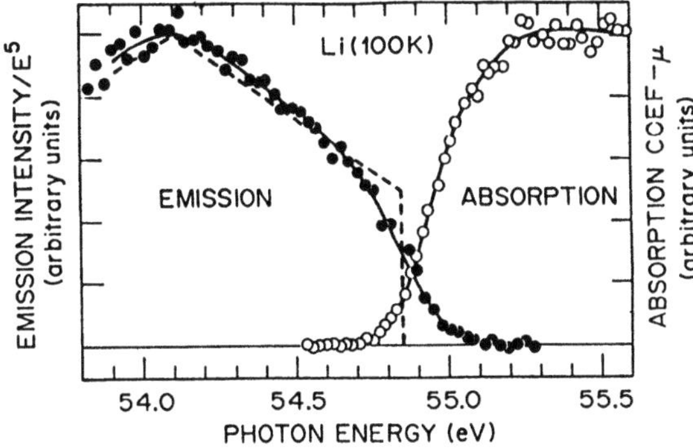

Fig.7 Emission spectra of lithium

Gaussian. This shape is always found in strong coupling situations. The dashed line to the left is the emission spectra for the case of long Auger lifetime (γ=0.0025 eV). This is the emission from the relaxed state, and shows the Franck-Condon shift of 2ζ= -12ω_0 for this model. The other dashed line shows the emission spectra for a short Auger lifetime (γ=0.05eV) where no relaxation occurs. Here one finds a broad gaussion which is slightly downshifted from the absorption peak. These two dashed lines characterize the extremes of no relaxation or complete relaxation.

Part (b) of Fig.6 shows the cases of incomplete relaxation. A double peaked spectra is found for the intermediate values of γ. For lithium the experimental value is γ=0.02 which has two nearly equal peaks. This broad function must be convoluted with the one-electron spectra, which results in the peak shape shown in Fig.7. This fits very well the experimental result of CALLCOTT et al.[27]. The broad emission edge of lithium, with its double shoulder, is caused by incomplete phonon relaxation around the core-hole.

There are two peaks in the lineshape function $E(\omega)$. One peak corresponds to emission from a relaxed state, while the other peak corresponds to emission from an unrelaxed state. When there are two peaks in $E(\omega)$ then some emission occurs from coreholes where relaxation has occurred, while other emission events take place from unrelaxed centers. This interpretation of the many-body phenomena suggests that emission does not occur from partially relaxed states.

At the same time this model was being proposed, similar mathematical models were being proposed in other contexts. SCHONHAMMER and GUNNARSSON [28] suggested that the two peaks observed in photoemission were from one electronic state in two possible stages of electronic relaxation. Their mathematical model is very similar to the one developed for IPR.

6. Experiments

Experiments on edge singularities began with the development at synchrotrons of light sources with continuous wavelengths. Edge singularities were observed in absorption experiments on simple metals. MND theory provided an explanation of why these singularities were observed in absorption from p-states but not from s-states. Examples are the experiments of ISHII et al.[29] on alkali metals, which are reproduced in

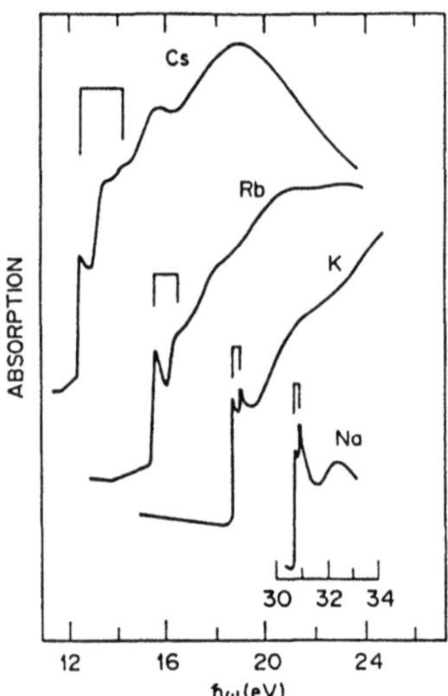

Fig. 8 X-ray absorption of alkali metals

Fig.8. There were also numerous experiments on lithium, sodium, magnesium and aluminum [3,27,30]. These early comparisons between experiment and theory were mainly qualitative.

MAHAN [31] was the first to try a detailed comparison between experiment and theory. Such comparisons comprise a complete theory of edge shapes which includes a number of factors. For absorption he included the following factors:

1) L_2 edge energy
2) L_3 edge energy
3) L_2 edge Auger width
4) L_3 edge Auger width
5) s- and d-fractions of edge intensity
6) s-and d-exponents α_ℓ
7) one electron-spectra
8) band width W.

This fitting procedure includes nine parameters. Most of these are not known beforehand nor are available by independent measurements. In spite of the large number of parameters, this list is not sufficient. In a subsequent comparison between theory and experiment, CITRIN et al.[30] noted that one should also include

9) Phonon broadening
10) spin-orbit scattering
11) spin-exchange scattering.

The phonon broadening is a temperature dependent gaussian, and is different from the temperature-independent Lorentzian broading due to Auger effects. Several of these parameters are very hard to determine. The separate s-and d-wave fractions of the one-electron edge intensity must be found by a band structure calculation. Each calculation gives different results, and these numbers are not yet well known. Similarly, the role of spin-exchange scattering is quite controversial. Each calculation of their effects gives different results [32-34].

One-electron theory predicts a 2:1 ratio between the intensities of the L_3 and L_2 absorption edges. Experiments in simple metals show that this ratio differs from 2:1. ONODERA [35] suggests this difference is due to spin scattering.

A quantitative fit of MND theory to the experiments requires a lot of careful work plus a complete one-electron band calculation. So far this type of analysis has only been done for lithium, sodium, magnesium, and aluminum where the theory and the experiments agree well.

Unfortunately the theory is applied to other cases with much less care. Usually the edge is fitted with a single exponent which is incorrect if the initial core level has $\ell > 0$. These types of analyses mean very little.

A number of emission experiments have been done on stoichiometric alloys of simple metals. The $L_{2,3}$ spectra generally show edge singularities. This phenomena is interpreted as a confirmation of the theory, although quantitative fits of the data are usually lacking.

FLYNN's group [36-38] has tested MND theory by doing numerous experiments on alkali metal alloys. Their experimental samples are made by evaporation at low temperatures, so the alloys are amorphous and unannealed. The alloys consist of a host alkali plus an impurity which is either another alkali, a rare gas atom, or a halide. In some cases their absorption spectra from the impurity p-shell show edge singularities in qualitative agreement with MND theory. However, in other cases they observed absorption spectra where the threshold is a ramp which would correspond to an exponent of $\alpha = -0.9$ or -1.0. These values of exponents are not permissible in MND theory. Flynn argues that his experiments repudiate the theory. His analysis of the data ignores a number of other explanations for his edge shape: (1) SUZUKI et al.[39] have suggested the importance of inhomogeneous broadening. Each impurity sees a different configuration of neighbors which gives a different random local field. These shift the core levels around, so that one has a distribution of threshold energies. This good idea has never been calculated in detail. (2) Conduction electron lifetime. In pure metals the lifetime is infinity at the Fermi energy. There is a finite lifetime in amorphous systems, which will broaden the edge singularities. No one has yet investigated whether MND theory is affected differently by elastic or inelastic scattering. (3) d-wave intensities. The edges should be substantially rounded if the d-wave intensity is high. One recent alloy calculation [40] shows about 50% d-wave intensity for krypton in rubidium. This calculation is shown in Fig.9. The linear line shape is from p-to-d transitions. We

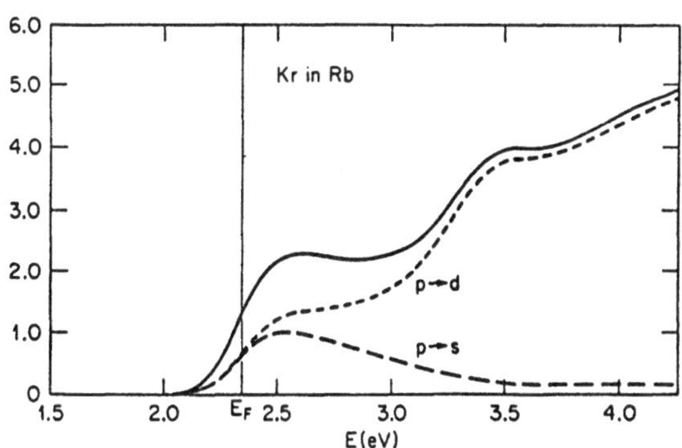

Fig.9. Theoretical krypton absorption in rubidium

summarize by emphasizing the importance of Flynn's experiments, and by
encouraging more work on these systems.

7. Dynamic Screening

The MND model has been solved exactly in terms of the dispersion integrals.
The most important remaining theoretical question is to understand the role
of dynamic screening in the absorption spectra. Theorists generally agree
that very near threshold, in the region of the edge singularities, it is a
good approximation to assume that the corehole is fully screened. Far from
the absorption threshold, in the EXAFS region, the corehole should be taken
as unscreened. The transition between these two regions is unclear.

FRIEDEL's group [41-42] has investigated dynamic screening in absorption
using a model where the excited electron leaves with constant velocity.
This classical treatment of the final state electron is okay if the
electron has a large kinetic energy on the order of kilovolts. This model
is fine for analysing XPS experiments, where we introduced it much earlier
for examining how the final electron creates excitation during its outward
trajectory [43]. However, the classical model is inappropriate for the
treatment of final state electrons of low kinetic energy. Here quantum
effects are important. This classical model cannot really solve the issue
of dynamic screening in absorption. Recent calculations by CANRIGHT and
MAHAN [44] have shown that the electron-hole pairs are an important part of
screening core holes. The time scale for the pair part of the formation of
the screening charge is still given by the plasma frequency.

The role of dynamic screening in the renormalization catastrophe has been
investigated by LANGRETH [45]. His model was also investigated by
MINNHAGEN [46]. Langreth suggested the formula for the kernal $R(u)$ in the
core hole Green's function (16)

$$R(u) = \frac{1}{u} \int \frac{d^3q}{(2\pi)^3} V_0(q)^2 (q^2/4\pi e^2) \, \text{Im}[\frac{1}{\epsilon(q,u)}] , \qquad (17)$$

where V_0 is the unscreened potential energy between the corehole and the
conduction electrons. Dynamic screening is included through the use of the
frequency-dependent dielectric function $\epsilon(q,\omega)$. This expression is not
exact, but is only the leading term in perturbation theory.

Langreth's expression could be improved by adding dispersion integrals.
Here we provide a _guess_ as to how they might enter the result. The
argument of the above integral can be rewritten as

$$\left| \frac{V_0(q)}{\epsilon(q,u)} \right|^2 \text{Im}\{P(q,u)\} , \qquad \text{where}$$

$$\text{Im}\{P(q,u)\} = \pi \int \frac{d^3k}{(2\pi)3} n_k (1-n_{k+q}) \left| t_h(\epsilon_k) t_e(\epsilon_{k+q}) \right|^2 \delta(u + \epsilon_k - \epsilon_{k+q}) . \qquad (18)$$

The dispersion integrals (14) for electrons and holes have been inserted
into the usual golden rule formula for the excitation of an electron of
initial wave vector k to an excited state above the Fermi energy with wave
vector k+q. We expect this formula to give reasonably good results for the
evaluation of dynamic screening on the corehole Green's function. No
similar formula exists for the role of dynamic screening on the exciton
part of the absorption spectra. Finding that formula is still a
theoretical challenge.

<u>8. Particle Hopping</u>

Singular phenomena occur whenever one adds or removes a charged particle from a localized state in a metal. The conduction electrons must adjust to this change in their environment, and this process is called screening. These effects are associated with the renormalization catastrophe.

Recently there has been much discussion of these effects associated with the hopping of localized particles in metals. One example is the diffusion of muons [47] and another is the motion of heavy fermions[48]. All of these theories bear a strong relationship to the classic treatment of HOLSTEIN [49] of large and small polarons. Holstein solved a model where the electron could hop from site to site. At each atomic site it was coupled to a set of local oscillators which are phonons. At very low temperatures the phonons provide a noise spectrum that weakly scatters the particle during its band motion. The resistivity increases with temperature since there are more phonons at higher temperature. Then at even higher temperatures the phonon noise localizes the particle. Its motion becomes diffusive as it hops from site to site. Then the phonons provide a heat bath which can thermally activate the hopping motion. Here the resistivity decreases with increasing temperature.

The same behavior is associated with the electronic screening of the diffusing particle. The screening charge can be represented by a set of bosons. The bosons are collective states such as plasmons or electron-hole pairs. A localized particle such as a muon or f-electron will have a strong coupling to the screening charge at each site. We can use the theory of SCHOTTE and SCHOTTE [50] who treated the renormalization catastrophe as a problem where the corehole is coupled to a set of electronic bosons. The same model is used to discuss the hopping of localized particles. Let $Q_q = (a_q + a_q^+)$ be the be displacement operator for phonons. The change in potential when a localized particle hops from site R to R+a is

$$\delta V = \sum_q M_q Q_q [e^{iq \cdot R} - e^{iq \cdot (R+a)}] .$$

The correlation function for particle hopping has an exponential factor of the type shown in (15) where

$$R(u) = \sum_q (M_q/\omega_q)^2 \left| 1 - e^{iq \cdot a} \right|^2$$

and ω_q is the boson energy. The quantity in brackets can usually be spherically averaged, which replaces it with

$$\left| 1 - e^{iq \cdot a} \right|^2 \Rightarrow 2[1 - j_0(2qa)] ,$$

where j_0 is the spherical Bessel's function. The above analysis is only valid for phonons or other ideal bosons. A good theory for the electronic screening of the electron gas is to use a formula for R(u) such as (17). Eqn.(17) is valid for a single particle appearing or disappearing at a site. The extension to a particle which hops a distance <u>a</u> is to insert into the integrand of (17) or (18) the factor of $2[1-j_0(2\overline{q}a)]$. At large <u>a</u> the Bessels function is small and there is just the factor of 2. Two is one plus one, where one factor is from the particle disappearing at one site while the other factor is from its appearance at another site. The Bessel's function provides interference when the value of <u>a</u> is small.

59

This discussion of the electronic polarization phenomena caused by a hopping particle is intended to be introductory. More rigorous treatments are provided in the references. TANABE and OHTAKA [51] solved it exactly for an MND type of model.

9. References

1. G.D. Mahan, Phys. Rev. 163, 612 (1967)
2. R. Haensel, G. Keitel, P. Schreiber, B. Sonntag, and C. Kunz, Phys. Rev. Lett. 23, 528 (1969)
3. T.A. Callcott, E.T. Arakawa, and D.L. Ederer, Phys. Rev. B 18, 6622 (1978)
4. G.D. Mahan, Phys. Rev. 153, 882 (1967)
5. P. Nozieres and C.T. DeDominicis, Phys. Rev. 178, 1097 (1969)
6. G.D. Mahan, Many-Particle Physics (Plenum, 1981) Sec. 8.3
7. J. Kondo, Prog. Theor. Phys. (Kyoto) 32, 37 (1964)
8. L.N. Cooper, Phys. Rev. 104, 1189 (1956)
9. M.S. Skolnick, J.M. Rorison, K.J. Nash, D.J. Mowbray, P.R. Tapster, S.J. Bass, and A.D. Pitt, Phys. Rev. Lett. 58, 2130 (1987)
10. M Combescot and P. Nozieres, J.Phys.(Paris) 32, 913 (1971)
11. P.W. Anderson, Phys. Rev. Lett. 18, 1049 (1967)
12. G.D. Mahan, Phys. Rev. B 21, 1421 (1980)
13. V.I. Grebennikov, Yu.A. Babanov, O.B. Sokolov, phys. stat. sol. (b) 79, 423 (1977); 80, 73 (1977)
14. U. von Barth and G. Grossmann, Phys. Rev. B 25, 5150 (1982)
15. L.C. Davis and L.A. Feldkamp, Phys. Rev. B 23, 4269 (1981)
16. T.A. Green, Phys. Rev. B 32, 3442 (1985)
17. W.J. Pardee and G.D. Mahan, Phys. Lett. 45A, 117 (1973)
18. D.R. Penn, S.M. Girvin, and G.D. Mahan, Phys. Rev. B 24, 6971 (1981)
19. G.D. Mahan, Phys. Rev. B 25, 5021 (1982)
20. K. Ohtaka and Y. Tanabe, Phys. Rev. B 28, 6833 (1983); B 30, 4235 (1980)
21. Y. Tanabe and K Ohtaka, Phys. Rev. B 29, 1653 (1984); B 32, 2036 (1985)
22. Y. Tanabe, Phys. Rev. B 33, 2806 (1986)
23. T. Kita, K. Ohtaka and Y. Tanabe, J. Phys. Soc. Jpn.
24. M.L. Goldberger and K.M. Watson, Collision Theory (Wiley, 1964) Appendix G
25. G.D. Mahan, Phys. Rev. B 15, 4587 (1977)
26. C.O. Almbladh, Solid State Comm. 22, 339 (1977)
27. T.A. Callcott, E.T. Arakawa, and D.L. Ederer, Phys. Rev. B 16, 5105, (1977)
28. K. Schonhammer and O. Gunnarsson, Solid State Comm. 23, 691 (1977); 26, 147,399 (1978); Phys. Rev. Lett. 41, 1608 (1978)
29. T. Ishii, Y. Sakisaka and S. Yamaguchi, J. Phys. Soc. Jpn. 42, 876 (1977)
30. P.H. Citrin, G.K. Wertheim, and Y. Baer, Phys. Rev. B 16, 4268 (1977); B 20, 3067 (1979)
31. G.D. Mahan, Phys. Rev. B 11, 4814 (1975)
32. S.M. Girvin and J.J. Hopfield, Phys. Rev. Lett. 34, 1320 (1975)
33. A. Yoshimori and A. Okiji, Phys. Rev. B 16, 3838 (1977)
34. H. Kaga, J. Phys. Soc. Jpn 43, 1144 (1977)
35. Y. Onodera, J. Phys. Soc. Jpn. 39, 1482 (1975)

36. R.A. Tilton, D.J. Phelps and C.P. Flynn, Phys. Rev. Lett.
    $\underline{32}$, 1006 (1974)
37. T.H. Chiu, D. Gibbs, J.E. Cunningham, and C.P. Flynn, Phys.
    Rev. B $\underline{32}$, 588 and 602 (1985)
38. C.P. Flynn, Surf. Sci. 158, 84 (1985)
39. S. Suzuki. T. Hanyu, H.Fukutani, H. Sugawara and I.
    Nagakura, S. Nakai, T. Ishii, H. Kato, and T. Miyahara,
    Solid State. Comm. $\underline{50}$, 769 (1984)
40. D.E. Meltzer, F.J. Pinski, and G.M. Stocks (preprint)
41. C. Nogura, D. Spanjaard, and J. Friedel, J. Phys. $\underline{F9}$, 1189
    (1979)
42. C. Nogura, in X-ray and Atomic Inner-Shell Physics-82, ed
    B. Crasemann (AIP, 1982) pg 676
43. G.D. Mahan, phys.stat.sol.(b) $\underline{55}$, 703 (1973)
44. G.S. Canright and G.D. Mahan (preprint)
45. D.C. Langreth, Phys. Rev. B 1, 471 (1970)
46. P. Minnhagen, J. Phys. F 7, (1977)
47. J. Kondo, Physica $\underline{125B}$, 279 (1984);126B, 377 (1984
48. S.H. Liu, Phys. Rev. Lett. $\underline{58}$, 2706 (1987)
49. T. Holstein, Adv. Phys. $\underline{8}$, 343 (1959)
50. K. Schotte and U. Schotte, Phys. Rev. 182, 479 (1969)
51. Y. Tanabe and K. Ohtaka, Phys. Rev. B $\underline{34}$, (1986)

Bethe Ansatz Treatment of the Anderson Model for a Single Impurity

A. Okiji

Department of Applied Physics, Osaka University, Suita,
Osaka 565, Japan

The exact solution of the Anderson model for a single impurity in metals is reviewed in this article.

1. Introduction

The Anderson Hamiltonian has been one of the most important Hamiltonians in the field of the solid state physics [1]. In fact this Hamiltonian has been adopted to explain the magnetic and electronic phenomena in dilute alloys including the Kondo effect, the valence-fluctuation phenomena in the rare-earth compounds [2] and the dynamical processes of the adsorbed atom on metal surfaces [3]. Therefore lots of theoretical studies have been done with the aim of investigating this model more precisely [2,4].

In the meantime, an exact solution of the Anderson model has been obtained with the use of the Bethe Ansatz method [5,6]. Consequently, the Kondo effect, which is one of the local many-body effects in solids, has been understood completely through the exact solution [5,6,7]. Furthermore, it is expected that using this fact will enable researchers to evaluate other local effects in solids with close connections with the Anderson orthogonal theorem.

In this article, the results obtained in the single orbital Anderson model are reviewed for the ground state properties and also for the thermodynamic properties in Chapter 2. The results of the orbitally degenerate Anderson model with the strong correlation are reported successively in Chapter 3. Some interesting things done so far as the application of the exact solution are mentioned in Chapter 4.

2. Exact solution of the single orbital Anderson model

2.1 Bethe-Yang-type equations

The Anderson Hamiltonian with a single localized orbital is [1]

$$H_A = \sum_{k,\sigma} \epsilon_k c^+_{k\sigma} c_{k\sigma} + \sum_{k,\sigma} V_k (c^+_{k\sigma} d_\sigma + d^+_\sigma c_{k\sigma}) + \epsilon_d \sum_\sigma d^+_\sigma d_\sigma + U d^+_\uparrow d_\uparrow d^+_\downarrow d_\downarrow \ . \qquad (2.1)$$

Here ϵ_k is the energy of a conduction-electron state; ϵ_d is the unperturbed energy of the d-state on an impurity atom; U is the Coulomb repulsive energy between d electrons; and V_k is the matrix element of the hybridization. Supposing that V_k does not depend on k and the values of U, $|\epsilon_d|$ and $|V|$ are small compared with the Fermi energy, the following deductions can be made: the system reduces to a one-dimensional one; the impurity effect of electron states far from the Fermi surface on the physical quantities can be neglected; then the energy spectrum, ε_k may be expressed as a linear function of k. The last statement means that the density of states for the conduction electron is constant and the band width is infinite. It is noted that these assumptions have been used in general. Eventually, the Anderson Hamiltonian can be rewritten in the following form without losing any essential and physical meaning [5].

$$H_A = \int dx[c_\sigma^+(x)(-id/dx)c_\sigma(x)+V\delta(x)\{c_\sigma^+(x)d_\sigma+d_\sigma^+c_\sigma(x)\}]$$
$$+ \varepsilon_d\sum_\sigma d_\sigma^+d_\sigma + Ud_\uparrow^+d_\uparrow d_\downarrow^+d_\downarrow \quad , \tag{2.2}$$

where the impurity is located at x=0 and the Fermi energy is chosen to be zero. This is the Anderson Hamiltonian written in the one-dimensional form.

Now, we can consider the application of the Bethe Ansatz method to this model. In the case of U being zero, the system becomes a simple one and is described by the simple product of the single electron wave function written by

$$|\psi_k\rangle = [\int g_k(x)c^+(x)dx+e_kd^+] \, |\,0\rangle \quad , \quad \text{where} \tag{2.3}$$

$$g_k(x) = e^{ikx}[1 - \frac{i}{2} \, \text{sgn}(x) \, \frac{V^2}{k-\varepsilon_d}] \quad ,$$

$$e_k = V/(k-\varepsilon_d) \quad ,$$

$$E = k \quad , \quad \text{and}$$

$|0\rangle$ means the vacuum state. The total energy is $E = 2\sum_1^{\frac{1}{2}N} k_i$, where N is total electron number. In the case where U is not equal to zero, we first consider two-electron problem of the singlet state. In that case, the wave function for two electrons is [5]

$$|\psi_1\rangle = |\psi_1\rangle + |\psi_2\rangle + |\psi_3\rangle \quad ,$$

$$|\psi_1\rangle = \int dx_1dx_2g(x_1,x_2)c_\uparrow^+(x_1)c_\downarrow^+(x_2)|0\rangle \quad ,$$

$$|\psi_2\rangle = \int dx \, e(x)[c_\uparrow^+(x)d_\downarrow^+-c_\downarrow^+(x)d_\uparrow^+]|0\rangle \quad ,$$

$$|\psi_3\rangle = f \, d_\uparrow^+d_\downarrow^+|0\rangle \quad , \quad \text{where} \tag{2.4}$$

$$g(x_1,x_2) = g_{k_1}(x_1)g_{k_2}(x_2) \, z(x_2-x_1) + g_{k_2}(x_1)g_{k_1}(x_2) \, z(x_1-x_2) \quad ,$$

$$e(x) = e_{k_1}g_{k_2}(x)z(x) + e_{k_2}g_{k_1}(x)z(-x) \quad ,$$

$$z(x) = 1+i\,\text{sgn}(x)UV^2/[k_1(k_1-U-2\varepsilon_d)-k_2(k_2-U-2\varepsilon_d)] \quad , \tag{2.5}$$

$$f = 2V/(k_1+k_2-U-2\varepsilon_d)$$

and the energy E is k_1+k_2. It is to be noted that $g(x_1,x_2)$ and $e(x)$ can be expressed by the single particle wave function and that U is included in the phase factor z.

Next, we consider the N-electron system. It is assumed that the wave function of N-electron system defined at the domain Q which is specified by $[x_{q_1}<x_{q_2}<\cdots<x_{q_N}]$ can be written in the following form [8]:

$$g^Q(x_1,x_2,\ldots,x_N) = \sum_{P=\{k_1 k_2 \cdots k_N\}} A(Q;P) \prod_{j=1}^{N} g_{k_{p_j}}(x_{q_j}) \quad ,$$

where the sum is over all states expressed by the permutation, P, **of the set** $\{k_i\}$. The phase coefficient $A(Q;P)$ is specified by the domain Q and the permutation $P=[k_1,k_2,\cdots,k_N]$ and the number of their values is $N!\times N!$.

What is the important thing is that we assume that the wave functions of N-electron system can be expressed by the product of the single particle wave functions mentioned above.

Here, we introduce the following operator Y_{ij}^{ab} for A:

$$Y_{ij}^{ab} A(\cdots, x_{qa},x_{qb}\cdots;\cdots,k_i,k_j\cdots)$$

$$= A(\cdots,x_{qa},x_{qb}\cdots \; ;\cdots,k_j,k_i\cdots) \quad .$$

It is necessary to satisfy the following conditions in order to obtain the self-consistent solutions [8]:

$$Y_{ij}^{ab} Y_{ji}^{ab} = 1 \quad , \qquad Y_{jk}^{ab} Y_{ik}^{bc} Y_{ij}^{ab} = Y_{ij}^{bc} Y_{ik}^{ab} Y_{jk}^{bc} \quad .$$

In other words, these are the conditions whether the Bethe Ansatz is applicable or not. It is easily deduced that the Y_{ij}^{ab} has to have the following form, in order to satisfy the above conditions[5]:

$$Y_{ij}^{ab} = \frac{-1+[f(k_i)-f(k_j)]P_{ab}}{1+[f(k_i)-f(k_j)]} \quad , \tag{2.6}$$

where $f(x)$ means arbitrary function and P_{ab} is the exchange operator of x_{qa} and x_{qb}. The important point is that this expression has the separable form about k_i and k_j. Furthermore we introduce the operator X_{ij}^{ab} as

$$X_{ij}^{ab} A(\cdots x_{qa},x_{qb},\cdots;\cdots k_i,k_j\cdots)$$

$$= A(\cdots x_{qb},x_{qa},\cdots;\cdots k_j,k_i\cdots) \; ,$$

which can be expressed by

$$X_{ij}^{ab} = P_{ab}\, Y_{ij}^{ab}$$

$$
= \frac{1-P_{ab}/[f(k_i)-f(k_j)]}{1+1/[f(k_i)-f(k_j)]} \quad . \tag{2.7}
$$

On the other hand, using the wave function for two electrons mentioned above, we can write down the explicit form of X_{ij}^{ab} for the Anderson model as

$$
X_{ij}^{ab} = \frac{1-iUV^2P_{ab}/[B(k_i)-B(k_j)]}{1+iUV^2/[B(k_i)-B(k_j)]} \quad , \tag{2.8}
$$

where $B(k)=k(k-U-2\epsilon_d)$. In the above expression, we include not only singlet but also triplet cases. The expression for X_{ij}^{ab} for the Anderson model has the same form as the general expression (2.7) which satisfies the applicable condition of the Bethe Ansatz method. That is, we can use the Bethe Ansatz method for the Anderson model.

The periodic boundary condition of this system written as

$$
g(x_1, \cdots, x_j+L, \cdots, x_N) = g(x_1, \cdots, x_j, \cdots, x_N)
$$

can be expressed in the following form:

$$
\frac{1-\frac{i}{2}V^2/(k-\epsilon_d)}{1+\frac{i}{2}V^2/(k-\epsilon_d)} e^{ik_jL} A(I,I) = X_{(j+1)j}X_{(j+2)j}\cdots X_{(j-2)j}X_{(j-1)j}A(I,I)
$$

with the use of $X_{ij} \equiv X_{ij}^{ij}$. Here L means the length of this system and I means the identity permutation. The factor

$\{1 - \frac{i}{2}V^2/(k-\epsilon_d)\}/\{1 + \frac{i}{2}V^2/(k-\epsilon_d)\}$ appears when an electron passes the impurity site. In order to represent the spin freedom, we have to use the spin-space operator X_{ij}^{σ} instead of real-space operator X_{ij}, which is given by

$$
X_{ij}^{\sigma} = \frac{1 + iUV^2P_{ij}^{\sigma}/[B(k_i)-B(k_j)]}{1 + iUV^2/[B(k_i)-B(k_j)]} \quad . \tag{2.9}
$$

Here, P_{ij}^{σ} is spin exchange operator and $P_{ij}^{\sigma}\cdot P_{ij}=-1$. Then the equations can be written in spin space and the Bethe Ansatz method can be applied to the problem expressed in spin space again, which gives rise to the diagonalized results of this system. Finally, one can obtain the following coupled equations, which were deduced by Wiegmann [9]:

$$
\exp[i(k_jL-\phi_j)] = \prod_{\beta=1}^{M} \frac{B(k_j)-\Lambda_\beta+iUV^2/2}{B(k_j)-\Lambda_\beta-iUV^2/2} \quad , \quad j = 1,2,\cdots,N, \tag{2.10a}
$$

$$- \prod_{j=1}^{N} \frac{B(k_j)-\Lambda_\alpha-iUV^2/2}{B(k_j)-\Lambda_\alpha+iUV^2/2} = \prod_{\beta=1}^{M} \frac{\Lambda_\alpha-\Lambda_\beta+iUV^2}{\Lambda_\alpha-\Lambda_\beta-iUV^2} \quad , \quad \alpha = 1,2,\cdots,M, \qquad (2.10b)$$

where $\phi_j = 2\tan^{-1}[V^2/2(k_j-\varepsilon_d)]$, $B(k) = k(k-U-2\varepsilon_d)$. Here, $N(M)$ is the total (up-spin) electron number, k_j denotes a quasimomentum and Λ_α is a variable related to the spin. Since the interaction depends on the internal freedom of spin at the impurity site through the symmetry of the wave function, the variable Λ_α is introduced in order to express this internal freedom and is determined by the second equations. It is to be noted that these equations include the bound state type solution of spin singlet as the ground state one and that, mathematically speaking, the solutions of these equations for the Anderson Hamiltonian are bit different from those for the s-d model with regard to this point. The solution of the s-d model will be mentioned again later.

2.2 Properties at the ground state

In general, the solutions of Eqs(2.10) describing the ground and excited states can be found in the complex plane and have been obtained by Kawakami and present author, and also by Wiegmann. We showed [10] from these equations that the solution for the singlet ground state can be expressed by the complex $k_\alpha^\pm$ which satisfy $B(k_\alpha^\pm) = \Lambda_\alpha \pm iUV^2/2$, where Λ_α is real, and the magnetic excitation can be expressed by real k. Then we describe the physical quantities at zero temperature with the use of the real Λ and k. In the thermodynamic limit, by introducing the distribution functions for the real Λ and k, Eqs(2.10) can be rewritten as the following formulae [10,11,12]:

$$\sigma(\Lambda)-\int_{-\infty}^{a} R(\Lambda-\Lambda')\sigma(\Lambda')d\Lambda'+ \int_{-\infty}^{b} Q[\Lambda-B(k)]\rho(k)dk = \sigma_s(\Lambda) \quad , \qquad (2.11a)$$

$$\rho(k)+B'(k)\int_{-\infty}^{b} R[B(k)-B(k')]\rho(k')dk'+B'(k)\int_{-\infty}^{a} Q[\Lambda-B(k)]\sigma(\Lambda)d\Lambda = \rho_s(k) \quad . \qquad (2.11b)$$

Here, $\Delta(x)$, $R(x)$ and $Q(x)$ are introduced as the Fourier transforms of $\exp(-V^2|\omega|/2+i\omega\varepsilon_d)$, $[1+\exp(UV^2|\omega|)]^{-1}$ and $(1/2)\mathrm{sech}(UV^2\omega/2)$, respectively and

$$\sigma_s(\Lambda) = \int_{-\infty}^{\infty} Q[\Lambda-B(k)][1/2\pi+\Delta(k)/L]dk \quad ,$$

$$\rho_s(k) = 1/2\pi+\Delta(k)/L+B'(k)\int_{-\infty}^{\infty} R[B(k)-B(k')][1/2\pi+\Delta(k')/L]dk' \quad ,$$

$$B'(k) = dB(k)/dk \quad .$$

Note that the width of the virtual bound state is $V^2/2$ and the density of states for the conduction electron band is $1/2\pi$ in this calculation. The distribution functions $\sigma(\Lambda)$ and $\rho(k)$ are divided into two terms, that is,

$$\sigma = \sigma^c + \sigma^i/L \quad \text{and} \quad \rho = \rho^c + \rho^i/L \quad ,$$

where σ^c and ρ^c express the distribution functions for the conduction electrons and σ^i and ρ^i those for the impurity parts. The parameters a and b appearing in the above expressions are considered to correspond to the Fermi levels of the excitation related to the particle number change and of the spin excitation respectively. For the given values of U, ε_d and the applied magnetic field H, they are determined by the energy minimum principle. The results are

$$\int_{-\infty}^{a} \sigma^c(\Lambda)d\Lambda = (U+2\varepsilon_d)/2\pi, \qquad \int_{-\infty}^{b} \rho^c(k)dk = H/2\pi \quad , \tag{2.12}$$

where $g\mu_B=1$. That is, one can obtain the set of equations to be solved and calculate the physical quantities at zero temperature. The magnetization and the number of localized electrons, for example, can be calculated by the following formulae:

$$s_z = \frac{1}{2}\int_{-\infty}^{b} \rho^i(k)dk \quad , \quad n_d = 1 - \int_{-\infty}^{a} \sigma^i(\Lambda)d\Lambda \quad . \tag{2.13}$$

In the magnetic field H, the ground state energy can be expressed as [10]

$$E^i = 2\text{Re}\int_{a}^{\infty} k_\alpha^+ \sigma^i(\Lambda_\alpha)d\Lambda_\alpha + \int_{-\infty}^{b} (k-H/2)\rho^i(k)dk \quad . \tag{2.14}$$

In particular, the ground state energy for the symmetric case ($U=-2\varepsilon_d$) in a zero field can be calculated analytically and the following expression is deduced [6]:

$$E_s^i = E_s^i(U=0) + (V^2/\pi)\text{Re}(1+iu)\log(1+iu)$$

$$- \frac{V^2}{2\pi}\int_{0}^{\infty}dx \int_{-\infty}^{\infty}dy \frac{\sqrt{x}\ \text{cosech}\pi(x+y^2)}{(y+\sqrt{u}/2)^2+1/4u} \quad , \tag{2.15}$$

where $u=U/V^2$. Using the above expression, we can obtain the perturbational results [13] for $u\ll1$ and the energy gain proportional to $T_K[\sim(UV^2/\pi^2)^{\frac{1}{2}}\exp(-\pi u/4)]$ for $u\gg1$ (s-d model) [14]. The explicit expression for the expansion by U is

$$E_s - E_s(U=0) = -\pi\Delta\{\frac{1}{4}\frac{U}{\pi\Delta} + 0.036861(\frac{U}{\pi\Delta})^2$$

$$-0.000795(\frac{U}{\pi\Delta})^4+0.000021(\frac{U}{\pi\Delta})^6\cdots\} \quad , \tag{2.16}$$

where $\Delta = V^2/2$ [13,15].

With the use of Eqs(2.11), the magnetic susceptibility defined by $\chi_m=ds_z/dH$ can be expressed in the following form [12]:

$$4\pi\chi_m = [C_a D_b \rho^i(b)/\rho^c(b) - C_b D_a \sigma^i(a)/\sigma^c(a)][C_a D_b - C_b D_a]^{-1} \quad , \quad \text{where} \quad (2.17)$$

$$C_x = [\partial/\partial x \int_{-\infty}^{a} \sigma^c(\Lambda)d\Lambda]/\sigma^c(a) \quad , \quad D_x = [\partial/\partial x \int_{-\infty}^{b} \rho^c(k)dk]/\rho^c(b) \quad , \quad x=a,b.$$

In the symmetric Anderson model at H=0, the magnetic susceptibility is written down explicitly [12].

$$\tilde{\chi}_m = \tilde{\chi}_p + \tilde{\chi}_k \quad ,$$

$$\tilde{\chi}_p = \int_{-\infty}^{\infty} \frac{(1+u^2-4ux^2)e^{-\pi x^2}}{(1+u^2-4ux^2)^2+16u^3x^2} \, dx \quad , \tag{2.18}$$

$$\tilde{\chi}_k = \pi(2\sqrt{u})^{-1}\exp[\pi(u-u^{-1})/4] \quad ,$$

where $\tilde{\chi}_m = \pi V^2 \chi_m$. In the expression of $\tilde{\chi}_m$, $\tilde{\chi}_k$ seems to have a singular character at U=0. So, at first, I think that this system may be singular at U=0. But it becomes clear that singular behaviour at U=0 is just apparent. In fact, Zlatic and Horvatic showed that using the above expression, $\tilde{\chi}_m$ can be rewritten in the following form [16]:

$$\tilde{\chi}_m = e^{\pi u/4}\int_{0}^{\infty} e^{-\pi x^2}\frac{\cos(\pi\sqrt{u}\ x/2)}{1 - ux^2} \, dx \quad .$$

This expression does not include the singularity at U=0. The results obtained by the expansion of this expression by U are in agreement with those obtained by the perturbational calculation done by Yosida and Yamada [13]. For large u, one can obtain the well-known magnetic susceptibility for s-d model, which is proportional to $1/T_K$, obtained by Ishii and Yosida [17].

The charge susceptibility defined as $\chi_c = -dn_d/d\varepsilon_d$ can be expressed as [12]

$$\pi\chi_c = [C_a D_b \sigma^i(a)/\sigma^c(a) - C_b D_a \rho^i(b)/\rho^c(b)][C_a D_b - C_b D_a]^{-1} \quad . \tag{2.19}$$

For the symmetric Anderson model at H=0 [12],

$$\tilde{\chi}_c = \int_{-\infty}^{\infty} \frac{(1+u^2+4ux^2)e^{-\pi x^2}}{(1+u^2+4ux^2)^2-16u^3x^2} \, dx \quad , \tag{2.20}$$

where $\tilde{\chi}_c = \pi V^2 \chi_c/4$. For u>>1 (s-d model), $\tilde{\chi}_c$ tends to zero because the charge fluctuation is depressed in this limit. The numerical results for the magnetic and charge susceptibilities in the asymmetric Anderson model are given in ref.[11].

2.3 Comparison with the Bethe Ansatz solution of the s-d model

The s-d Hamiltonian is

$$H = \sum_{k,\sigma} \varepsilon_k C^+_{k\sigma} C_{k\sigma} + \sum_{k,k'} J_{kk'} [(C^+_{k'\uparrow} C_{k\uparrow} - C^+_{k'\downarrow} C_{k\downarrow}) S_z + C^+_{k'\uparrow} C_{k\downarrow} S^- + C^+_{k'\downarrow} C_{k\uparrow} S^+] \quad ,$$

$$(2.21)$$

where $J_{kk'}$ is the exchange interaction between the conduction electron and the localized electron with the spin S at the impurity site [14]. Here, we consider the case of J>0 (antiferromagnetic coupling) and S=1/2. This model was also solved exactly with the use of the Bethe Ansatz method and the thermodynamic properties were calculated precisely, which are reviewed in ref.[7]. The assumption for getting the exact solution of the s-d model with the use of the Bethe Ansatz method is essentially the same as that of the Anderson model. That is, J is independent of k and ε_k is proportional to k, which corresponds to the case where the band width of the conduction electrons is infinite. Note that one can derive the s-d Hamiltonian from the symmetric Anderson Hamiltonian (U=-2ε_d) approximately by bringing the value of U into infinity. In that case J=4V²/U [18]. The Bethe-Yang-type equations obtained for the s-d model are

$$e^{ik_jL} = \prod_{\beta=1}^{M} \frac{-\lambda_\beta + iJ/8}{-\lambda_\beta - iJ/8} \cdot \frac{1+iJ/4}{1-iJ/4} \qquad (2.22a)$$

$$-\left[\frac{-\lambda_\alpha - iJ/8}{-\lambda_\alpha + iJ/8}\right]^{N_c} \left[\frac{\tfrac{1}{4}-\lambda_\alpha - iJ/8}{\tfrac{1}{4}-\lambda_\alpha + iJ/8}\right] = \prod_{\beta=1}^{M} \frac{\lambda_\alpha - \lambda_\beta + iJ/4}{\lambda_\alpha - \lambda_\beta - iJ/4} \qquad , \qquad (2.22b)$$

where N_c is the number of the conduction electrons [19].

On the other hand, the Bethe-Yang-type equations for the symmetric Anderson model can be expressed as

$$e^{ik_jL} = \prod_{\beta=1}^{M} \frac{k_j^2 - \Lambda_\beta + iUV^2/2}{k_j^2 - \Lambda_\beta - iUV^2/2} \cdot \frac{k_j + \tfrac{1}{2}U + iV^2/2}{k_j + \tfrac{1}{2}U - iV^2/2} \qquad (2.23a)$$

$$-\prod_{j=1}^{N} \frac{k_j^2 - \Lambda_\alpha - iUV^2/2}{k_j^2 - \Lambda_\alpha + iUV^2/2} = \prod_{\beta=1}^{M} \frac{\Lambda_\alpha - \Lambda_\beta + iUV^2}{\Lambda_\alpha - \Lambda_\beta - iUV^2} \qquad , \qquad (2.23b)$$

which can be easily derived by substituting U=-2ε_d in Eqs(2.10). From a standpoint of the Anderson model it is considered to assume that in the derivation of the Bethe-Yang-type equations for the s-d model one of the quasimomenta corresponding to impurity level is $-\tfrac{1}{2}U$ and all other quasimomenta for the conduction electrons are much smaller than U. Accordingly, neglecting all of the k_j belonging to the right hand side of Eq(2.23a) of the symmetric Anderson model, one can obtain Eq(2.22a) of the s-d model. Furthermore, if one does the same operation for Eq(2.23b) except for one of the k_j which is corresponding to impurity electron ($k_j=-\tfrac{1}{2}U$), one can obtain Eq(2.22b) of the s-d model. It is clear that to find the bound-state-type solution in Eqs(2.22) is impossible for the ground state, and that the ground state energy and also the Kondo temperature, for example, cannot be calculated precisely with the use of the exact solution of the s-d model although the physical quantities scaled by the conventional T_K can be calculated exactly [7]. Eventually, in order to obtain the full information about the s-d model exactly, it is necessary to use the s-d limit of the exact solution of the Anderson model.

2.4 Thermodynamic properties

In order to investigate the thermodynamic properties of the Anderson model, we must seek all the solutions of Eqs(2.10) in the complex plane. In the thermodynamic limit, these solutions are grouped into the complex Λ, the complex k and the real k. The solution of the complex Λ has the following form [20]:

$$\Lambda_\alpha^{n,\ell} = \Lambda_\alpha^n + i(n+1-2\ell)UV^2/2 \quad , \quad \ell=1,2,\cdots,n, \quad n=1,2,\cdots,\infty$$

where Λ_α^n is real and the sets of $\Lambda_\alpha^{n,\ell}$ are the nth-order strings [21]. The solution of the complex k relates to $\Lambda_\alpha^{n,\ell}$ and can be written as

$$B(k_\alpha^{n,\ell}) = \Lambda_\alpha^{n,\ell} \pm iUV^2/2 \quad .$$

The distribution functions for these solutions should be determined by minimizing the thermodynamic potential. Here, we introduce the three kind of pseudoenergies instead of the distribution functions for the corresponding three kinds of solutions, for example, the pseudoenergy $\kappa = T\ln\rho^h(k)/\rho(k)$, where $\rho^h(k)$ and $\rho(k)$ express the distribution functions of the real k for the hole and the particle. These pseudoenergies are determined by the following set of the integral equations [20,22-24]:

$$\kappa(k) = \kappa_s + T\int_{-\infty}^{\infty} Q(B(k)-\Lambda)[G(\varepsilon_1^+)-G(\varepsilon_1^-)]d\Lambda \quad ,$$

$$\varepsilon_1^\pm(\Lambda) = TQ*G(\varepsilon_2^\pm) - T\int_{-\infty}^{\infty} B'(k)Q[B(k)-\Lambda]G(\pm\kappa)dk \quad ,$$

$$\varepsilon_n^\pm(\Lambda) = TQ*[G(\varepsilon_{n-1}^\pm)+G(\varepsilon_{n+1}^\pm)] \quad , \quad n>2 \quad ,$$

$$\lim_{n\to\infty} \varepsilon_n^+/n = U+2\varepsilon_d \quad , \quad \lim_{n\to\infty} \varepsilon_n^-/n = H \quad , \tag{2.24}$$

where $*$ means convolution and

$$G(x) = \log[1+\exp(x/T)] \quad ,$$

$$\kappa_s(k) = k - \tfrac{1}{2}U - \varepsilon_d + 2\int_{-\infty}^{\infty} (x - \tfrac{1}{2}U-\varepsilon_d)^2 R[B(k)-B(x)]dx \quad .$$

The expression for the impurity contribution to the thermodynamic potential can be expressed by [20,22-24]

$$\Omega^i = E_s^i - T\int_{-\infty}^{\infty} \sigma_s^i G(\varepsilon_1^+)d\Lambda - T\int_{-\infty}^{\infty} \rho_s^i G(-\kappa)dk \quad , \tag{2.25}$$

where $\sigma_s^i(\rho_s^i)$ is given by the impurity part of $\sigma_s(\rho_s)$ (see Eqs(2.11)).
At low temperatures the physical quantities can be described with the use
of the expression at zero temperature. For example, the calculation of the
thermodynamic potential up to T^2 gives the expression for the coefficient
of the T-linear specific heat [20,22]

$$\tilde{\gamma} = (V^2/8)[\sigma^i(a)/\sigma^C(a)+\rho^i(b)/\rho^C(b)] \quad , \tag{2.26}$$

where $\tilde{\gamma}$ is normalized by the value at $U=H=\varepsilon_d=0$. Using the Eqs(2.17),(2.19)
and (2.26), we can derive the following important relation exactly between
the coefficient of the T-linear specific heat and the susceptibilities at
$T=0$ [20,22]:

$$\tilde{\gamma} = (\tilde{\chi}_m+\tilde{\chi}_c)/2 \quad . \tag{2.27}$$

Note that the above relation is derived for arbitrary values of U, ε_d and
H. As for the symmetric Anderson model, this relation has been obtained by
the perturbational method [13]. This simple relation shows that $\tilde{\chi}_m$ and $\tilde{\chi}_c$
are the basic quantities which characterize the low temperature
thermodynamics of the Anderson model. In Fig.1, the numerical results of
$\tilde{\chi}_m$, $\tilde{\gamma}$, $\tilde{\chi}_c$ and also $R=\tilde{\chi}_m/\tilde{\gamma}$ are shown for the symmetric Anderson model as a
function of U [20]. Note that so-called Wilson's ratio R increases from
unity to 2 rather rapidly as U is increased, which means that the many-
body character appears at the rather small value of U. The results of $\tilde{\gamma}$
for the asymmetric case in an arbitrary value of H are given in Fig.2
for reference [22].

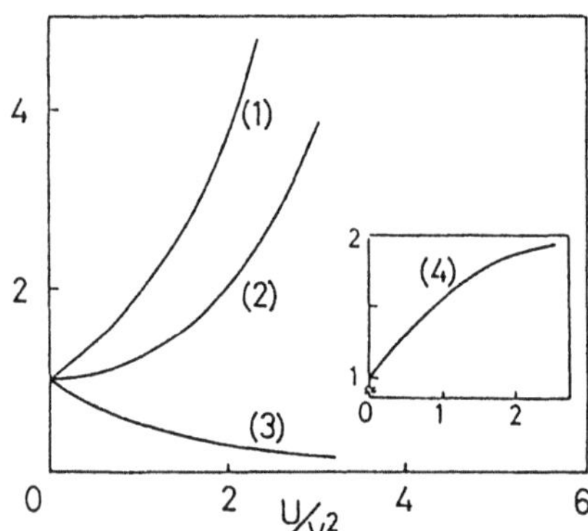

Fig.1. Curves (1),(2),(3) and (4)
represent $\tilde{\chi}_m$, $\tilde{\gamma}$, $\tilde{\chi}_c$ and $\tilde{\chi}_m/\tilde{\gamma}$ as
a function of U/V^2 in the symmetric
case.

The numerical calculation at finite temperatures is carried out by
truncating the integral equations (2.24) at finite numbers n and by
converting the integral equations into a set of matrix equations. These
matrix equations are iterated until the results attain stability. The
numerical results are thought to be accurate to within a few percent.
The numerical results for the magnetic susceptibility and the specific
heat over a physical relevant temperature range are shown in Figs.3,4
for the symmetric Anderson model [23] and in Figs.5,6 for the asymmetric

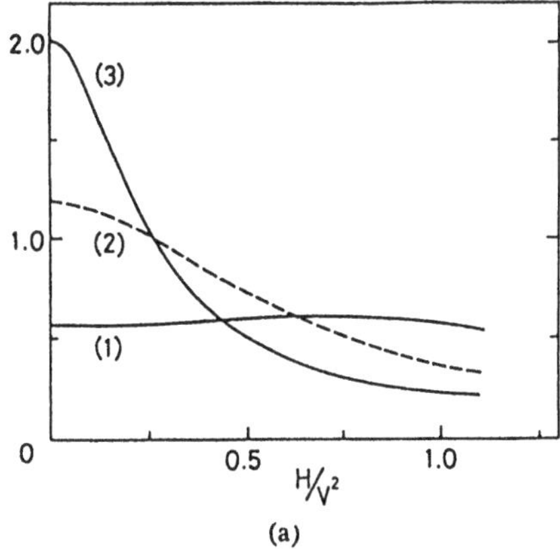
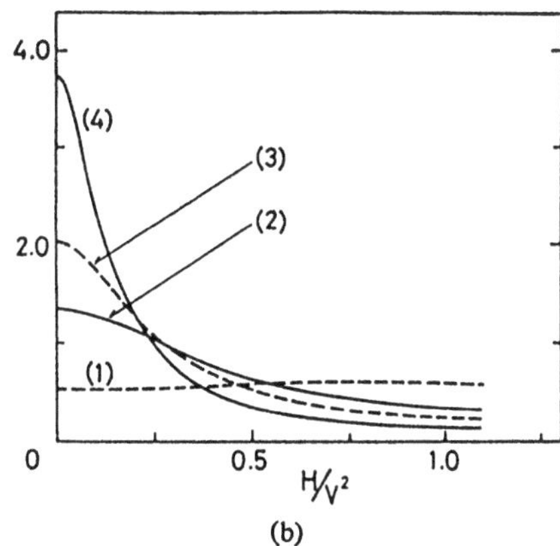

Fig.2. The curves of $\tilde{\gamma}$ as a function of H/V^2. (a) $U/V^2=2.0$: (1) $\varepsilon_d/V^2=0.12$ and -2.12, (2) -0.35 and -1.65, (3) -1.0 (b) $\varepsilon_d/V^2=-1.0$: (1) $U/V^2=0.7$, (2) 1.5, (3) 2.0, (4) 6.5

case [24]. The effects of a magnetic field on the specific heat are shown in Figs.7,8 [23,24].

In Fig.3 (curves 2,3) and Fig.5 (curve 1), the effective Curie constant $T\chi_m$ has a maximum value where a localized moment is considered to be developed. The localized moment is, however, quenched due to the correlation effect of U in the low temperature region, leading to the singlet bound state at zero temperature [14]. Correspondingly the specific heat has a peak structure in the low temperature region. In the symmetric case, the magnetic susceptibility takes the value $\chi_m=1/2\pi T_K$ at $T=0$, where the Kondo temperature T_K is $(UV^2/\pi^2)^{\frac{1}{2}}\exp[-(\pi/4)(U/V^2-V^2/U)]$ in this calculation. Note that the whole nature of the results for the magnetic susceptibility shows close agreement with the results of the renormalization-group calculation [7]. From Figs.7 and 8, one can see that the many-body singlet state caused by

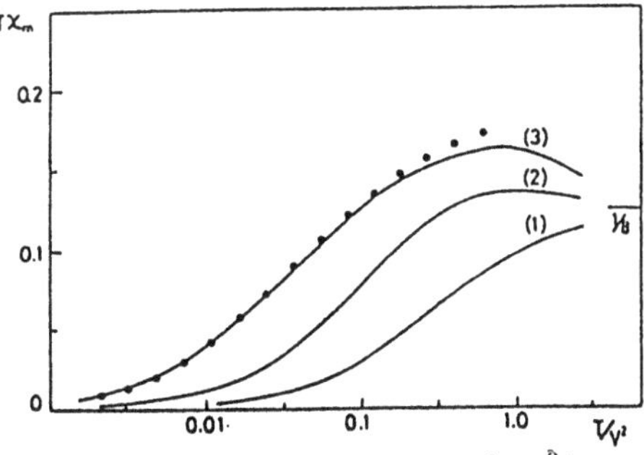

Fig.3. Plots of $T\chi_m$ vs $\log(T/V^2)$ for the symmetric cases where (1) $U/V^2=0$, (2) 2.0 and (3) 4.0. The points are the renormalization-group calculation results for the s-d model.

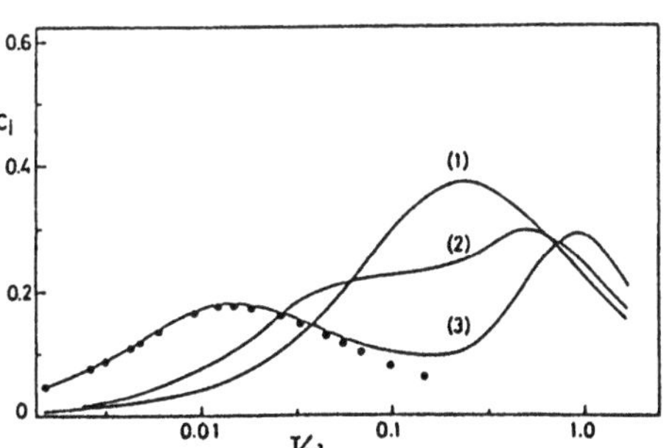

Fig.4. Plots of specific heat C_i vs $\log(T/V^2)$ for the symmetric cases where (1) $U/V^2=0$, (2) 2.0 and (3) 4.0. The points are the values of the exact calculation of the s-d model.

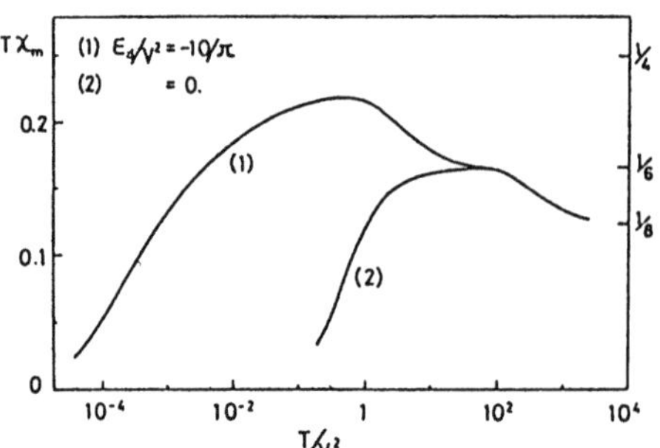

Fig.5. Plots of $T\chi_m$ vs $\log(T/V^2)$ for $U/V^2=10^3/\pi$. The values of ε_d/V^2 are (1) $-10/\pi$ and (2) 0.

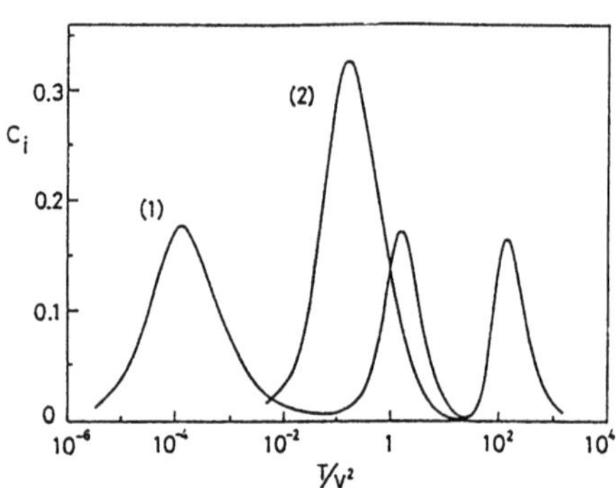

Fig. 6. Plots of specific heat C_i vs $\log(T/V^2)$ for $U/V^2=10^3/\pi$. The values of ε_d/V^2 are (1) $-10/\pi$ and (2) 0.

the strong correlation at the impurity site [14] is destroyed by the magnetic field as it is by the temperature [23,24].

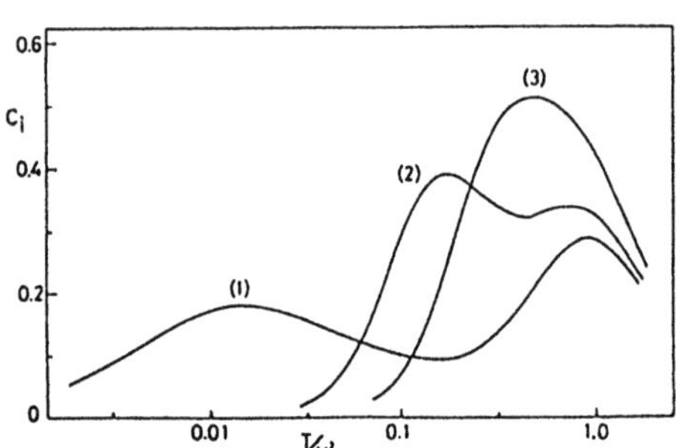

Fig.7. Plots of specific heat C_i vs $\log(T/V^2)$ for the symmetric case $U/V^2=4.0$ and the magnetic field (1) $H/V^2=0$ (2) 0.5 and (3) 1.0.

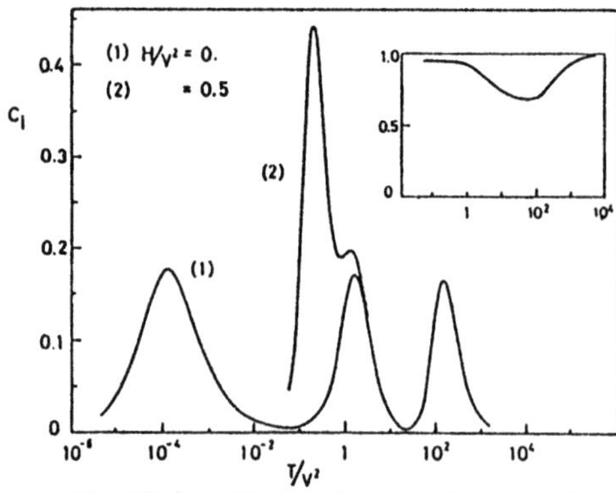

Fig.8(a). Plots of specific heat C_i vs $\log(T/V^2)$ for $U/V^2=10^3/\pi$ and $\varepsilon_d/V^2=-10/\pi$. The impurity occupancy $n_d(T)$ for $H=0$ is plotted vs $\log(T/V^2)$ in the inset.

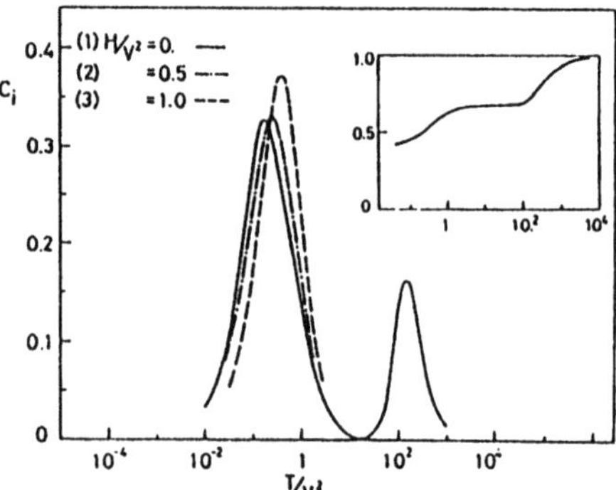

Fig.8(b). Plots of specific heat C_i vs $\log(T/V^2)$ for $U/V^2=10^3/\pi$ and $\varepsilon_d=0$. The impurity occupancy $n_d(T)$ for $H=0$ is plotted vs $\log(T/V^2)$ in the inset.

74

3. Bethe Ansatz treatment of the highly correlated degenerate Anderson model

3.1 Basic equations

Bethe Ansatz method for the single orbital Anderson model has been extended
to the more realistic cases, the Coqblin-Schrieffer model [25] and the
highly-correlated degenerate Anderson model which includes the Coqblin-
Schrieffer model as a special case [26]. These models have been intensively
studied in order to understand the properties of the dilute impurity systems
such as Ce (Yb) impurities in a metal and to clarify the local properties of
the valence-fluctuation phenomena in the concentrated systems. In the
calculation of the orbital-degenerate models, the effect of the degeneracy
has been found to play an important role at low temperatures [25-32]. For
example, the temperature-dependent susceptibility has a maximum at a finite
temperature, which did not appear in the original Kondo problem, and the
peak structure of the susceptibility curve becomes more prominent with the
increase of the degeneracy [28,31-33]. These results have explained some
experimental findings of Ce (Yb) compounds rather successfully [32,33].
The crystalline field which can be included in the calculation has also
come into play and, if it is large, affected the thermodynamic properties
considerably[34-36]. Note that the investigation of the crystalline field
effect in this problem has been done already with other theoretical
methods, perturbation [37-39] and scaling [40]. In this chapter, the
results obtained by means of the Bethe Ansatz method are reviewed, for the
thermodynamic properties of the highly-correlated degenerate Anderson
Hamiltonian with the crystalline field and the spin-orbit coupling [35,36].

At first, we summarize the procedure of the diagonalization briefly in
order to write down the basic algebraic equations. The Hamiltonian is
expressed as

$$H_T = H_A + H_W \quad , \tag{3.1a}$$

$$H_A = \sum_{k,m} \varepsilon_k C^+_{km} C_{km} + V \sum_{k,m} [C^+_{km} C_{fm} + C^+_{fm} C_{km}]$$

$$+ \varepsilon_f \sum_m C^+_{fm} C_{fm} + U \sum_{m \neq m'} C^+_{fm} C_{fm} C^+_{fm'} C_{fm'} \quad , \tag{3.1b}$$

and $U \to \infty$, where C^+_{fm} ($1 \leq m \leq \nu$) is the creation operator for the ν-fold degenerate
electron states of the impurity atom and C^+_{km} is for the conduction electron
in the partial wave representation. The second term H_W in H_T represents the
sum of the additional contributions from the magnetic field, the crystalline
field and the spin-orbit coupling, which may cause the splitting of the
ν-fold degenerate states in different ways. First, we consider the case
of V being zero. In that case we may write down the eigenvalue of H_W
explicitly as the form of H_W is known. We denote the eigenvalue of H_W,
$W_i(i=1\cdots\nu)$, in the following way:

$$W_1 \leq W_2 \leq \cdots \leq W_\nu \quad , \quad \sum_{m=1}^{\nu} W_m = 0 \quad . \tag{3.2}$$

It is to be noted that one should use the above index m specified by W_m by
way of the index m appearing in the Hamiltonian H_T if one intends to
consider the additional field H_W. The Bethe Ansatz diagonalization of H_T
is carried out in two steps in this case too with the same condition for the

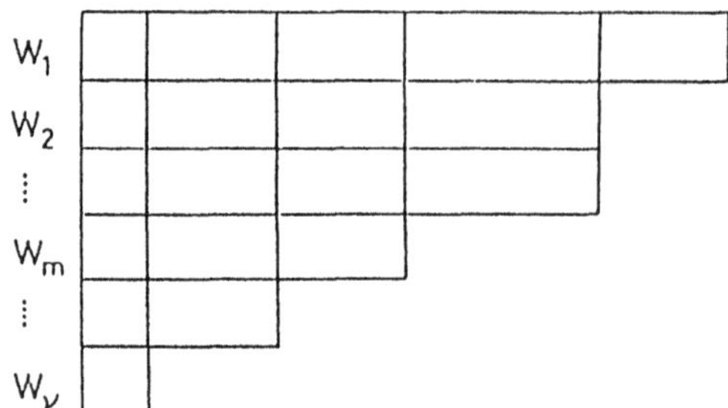

Fig.9. Young's tableau in the "spin space".

conduction electrons used in the preceding chapter. In the first step, we look for a solution of H_T as a superposition of plane waves characterized by the quasimomentum. $k_p^{(1)}$ (p=1,2,$\cdots$,N$_1$: N$_1$ is a number of total electrons) [8]. After applying the periodic boundary condition, the problem is rewritten as the ancillary problem within the spin space in which the permutation symmetry is characterized by Young's tableau with ν-rows (Fig.9).

In the present system the particle-state belonging to the m-th row should be considered to correspond to the state specified by W_m. In order to complete the diagonalization of the ancillary problem with the generalized Bethe Ansatz [8,41], we necessarily introduce the additional quasimomentum $k_p^{(j)}$ ($2 \le j \le \nu$). Eventually, the basic algebraic equations (Bethe-Yang type) for $k_p^{(j)}$ can be expressed in the following form [29,34,35]:

$$\exp(ik_p^{(1)}L)t(k_p^{(1)}-\varepsilon_f) = \prod_{q=1}^{N_2} t(k_p^{(1)}-k_q^{(2)}) \quad , \tag{3.3a}$$

$$-\prod_{q=1}^{N_j} t\{(k_p^{(j)}-k_q^{(j)})/2\} = \prod_{r=1}^{N_{j-1}} t(k_p^{(j)}-k_r^{(j-1)})\prod_{s=1}^{N_{j+1}} t(k_p^{(j)}-k_s^{(j+1)}) \quad ,$$
$$(2 \le j \le \nu) \tag{3.3b}$$

where $t(X)=(X-i\Delta)/(X+i\Delta)$, $\Delta(=V^2/2)$ is the resonance width of the virtual bound-state. Here, N_j-N_{j+1} ($N_{\nu+1}=0$) is the number of electrons belonging to the j-th row specified by W_j in Young's tableau.

In order to calculate the thermodynamic quantities described by the Hamiltonian (3.1a), we should seek all the possible solutions of Eqs.(3.3). The solution is divided into two kinds. They are,

(a) n-th order charge complex $k_q^{(j)}$ which makes bound states with the other solutions within the same column of Young's tableau:
$$k_q^{(j)} = k_q^{(j+1)} \pm i\Delta \quad ,$$
where $1 \le j \le n-1$, $1 \le n \le \nu$ and $k_q^{(n)}$ is real,

(b) m-th order spin complex $k_s^{(j)}$ which makes bound states within the same row of Young's tableau:
$$k_s^{(j)} = k_{sm}^{(j)} + i(m+1-2\ell)\Delta \quad ,$$

where $2 \le j \le \nu$, $1 \le \ell \le m$ (m=1,2,$\cdots$) and k_{sm} is real. Here, we introduce the distribution functions for the real part of the above solutions in the

76

thermodymamic limit: $\rho_0^{(j)}(\tilde\rho_0^{(j)})$ for the particle (hole) of the j-th order charge complex and $\rho_m^{(j)}(\tilde\rho_m^{(j)})$ for the particle (hole) of the m-th order spin complex. These distribution functions should be determined by minimizing the thermodynamic potential. The resulting equations are expressed as follows in **terms** of the pseudoenergies defined by
$\varepsilon_0^{(j)} = T\ln(\tilde\rho_0^{(j)}/\rho_0^{(j)})$ for $1\le j\le\nu$ and $\varepsilon_n^{(j)} = -T\ln(\tilde\rho_n^{(j+1)}/\rho_n^{(j+1)})$ for $1\le j\le\nu-1$, $1\le n$:

$$\varepsilon_0^{(\nu)}/T = k/T + R*G(\varepsilon_0^{(\nu)}) + \sum_{j=1}^{\nu-1} P_j*G(-\varepsilon_0^{(j)}) \quad , \tag{3.4a}$$

$$G(\varepsilon_n^{(j)}) = P_j*G(\varepsilon_0^{(\nu)})\delta_{no} + \sum_{\ell=1}^{\nu-1}\{V_{j\ell}*G(-\varepsilon_n^{(\ell)}) - U_{j\ell}*[G(-\varepsilon_{n+1}^{(\ell)}) + G(-\varepsilon_{n-1}^{(\ell)})]\}$$

$$\text{for } 1\le j\le\nu-1 \quad , \quad n=0,1,2,\cdots, \tag{3.4b}$$

$$\lim_{n\to\infty}\varepsilon_n^{(j)}/n = W_j - W_{j+1} \quad , \tag{3.4c}$$

where $*$ means convolution, T represents the temperature ($k_B=1$) and $G(x)=\log(1+\exp(x/T))$. Here P_j and R are functions defined by $\sinh(j\Delta\omega)/\sinh(\nu\Delta\omega)$ and $\exp(-\Delta|\omega|)\sinh[(\nu-1)\Delta\omega]/\sinh(\nu\Delta\omega)$ in Fourier space and $U_{j\ell}$ and $V_{j\ell}$ are given by

$$U_{j\ell} = \sum_{i=1}^{z} P_{|j-\ell|+2i-1} \quad ,$$

$$V_{j\ell} = \sum_{i=1}^{z} (P_{|j-\ell|+2i} + P_{|j-\ell|+2i-2}) \quad \text{and}$$

$$z = \min(j,\ell).$$

The impurity contribution to the thermodynamic potential is also expressed in terms of the pseudoenergies as

$$\Omega^i = -T \sum_{j=1}^{\nu} \int_{-\infty}^{\infty} \phi_j(k)G(-\varepsilon_0^{(j)})dk \quad , \quad \text{where} \tag{3.5}$$

$$\phi_j(k) = (j\Delta/\pi)/\{(k-\varepsilon_f)^2+(j\Delta)^2\} \quad .$$

With the use of the ancillary equations (3.4), the thermodynamic quantities of the Hamiltonian (3.1) can be calculated exactly if the explicit form of H_W is given and the corresponding eigenvalue of W_j is obtained explicitly [35]. It is noted that the effect of the additional field, H_W, appears only in the boundary condition through the value of W_j. This simple result is of much help for the further analytical and numerical calculations. It should also be noticed that an ultraviolet cut-off is necessarily introduced for $\varepsilon_0^{(\nu)}$ in this model, if one intends to calculate the physical quantities explicitly [31,32].

At low temperatures Eqs(3.4) can be treated analytically and the general expression for the T-linear specific heat can be written down in the explicit form. We can expand the pseudoenergy in Eqs(3.4) as $\varepsilon_0^{(j)} \sim n_j(k) + T^2 \lambda_j(k)$. Note that the contribution from ε_n $(n \geq 1)$ to the thermodynamic potential can be neglected within the accuracy of T^2. Then the expression for Ω^i up to T^2 has the form

$$\Omega^i = \Omega^i(T=0) + T^2 \sum_{j=1}^{\nu} \{ \int_{-\infty}^{Q_j} \phi_j \lambda_j dk - \frac{\pi^2}{6} \phi_j(Q_j)/(d n_j/dQ_j) \} \quad , \tag{3.6}$$

where the cut-off parameter Q_j is defined by $n_j(Q_j)=0$, which corresponds to the Fermi-level of each elementary excitation. The formula (3.6) can be rewritten in a simpler form in terms of the zero-temperature distribution functions $\rho_I^{(j)}$ and $\rho_C^{(j)}$. Consequently, the expression for the coefficient of T-linear specific heat is written down as [35]

$$\gamma = \frac{\pi}{6} \sum_{j=1}^{\nu} \{ \rho_I^{(j)}(Q_j)/\rho_C^{(j)}(Q_j) \} \quad , \tag{3.7a}$$

$$\sum_{\ell=1}^{j} \sum_{j=\ell}^{\nu} [i+j-2\ell]_i^* \, \rho_I^{(i)} = \phi_j(k) \quad , \tag{3.7b}$$

$$\sum_{\ell=1}^{j} \sum_{i=\ell}^{\nu} [i+j-2\ell]_i^* \, \rho_C^{(i)} = \frac{j}{2\pi} \quad , \quad \text{for } 1 \leq j \leq \nu \quad , \quad \text{where} \tag{3.7c}$$

$$[n]_i^* f = \int_{-\infty}^{Q_i} dx \, \frac{(n\Delta/\pi)f(x)}{(k-x)^2+(n\Delta)^2} \quad , \quad [0]_i^* f = f \quad .$$

The Fermi level, Q_j, is determined by

$$\int_{-\infty}^{Q_j} \rho_C^{(j)} dk = (W_{j+1}-W_j)/2 \quad , \quad \text{for } 1 \leq j \leq \nu-1 \quad , \tag{3.8a}$$

$$\sum_{j=1}^{\nu} \int_{-\infty}^{Q_j} j\rho_C^{(j)} dk = N_1/L \quad . \tag{3.8b}$$

The above expression for γ also shows that the low-temperature behaviour of this system is described by the local Fermi-liquid picture [7,13,33,37-40].

3.2 Effect of the orbital degeneracy on the susceptibility

At first, we mention the obtained results without the effect of a crystalline field as an example. The T-linear specific heat C_i for the ground multiplet of the impurity with the total angular momentum J can be expressed as [30]

$$C_i/T = \frac{2\pi^2}{(J+1)(2J+1)} \chi_m + \frac{\pi^2}{3(2J+1)} \chi_c \quad , \tag{3.9}$$

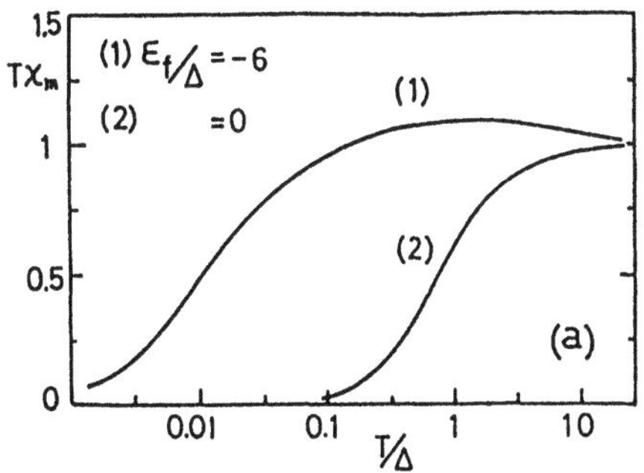
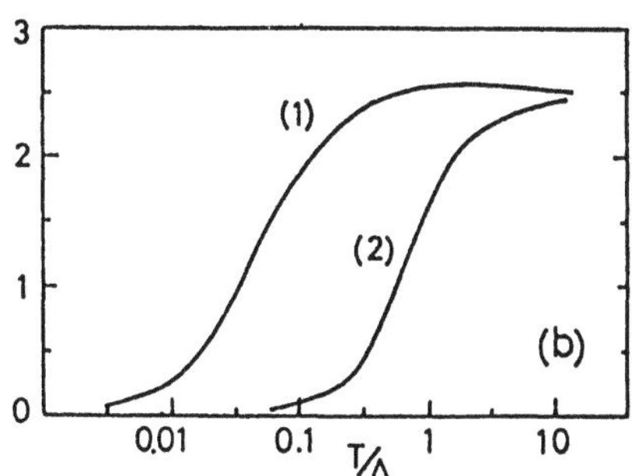

Fig.10. Plots of the effective Curie constant TX_m vs $\log(T/\Delta)$ for (a) $J=3/2$ and (b) $J=5/2$.

where X_m and X_c are the magnetic and charge susceptibilities at zero temperature [29,42]. This relation is considered as a generalized local-Fermi-liquid relation for arbitrary J and as a natural extension of the expression (2.27). If ε_f (the energy of a localized electron) lies far above the Fermi level, the value of the ratio $R = TX_m/C_i$ tends to $J(J+1)/\pi^2$ [33,43] and if ε_f lies far below the Fermi level, R tends to a generalized Wilson's ratio $(J+1)(2J+1)/2\pi^2$ [40]. The numerical results of the effective Curie constant TX_m are shown in Fig.10.

We can say that the general aspect of TX_m is independent of J and is essentially the same as that for the $J=1/2$ case [4,23,24,28]. The effect of degeneracy, however, appears rather significantly in the behaviour of the susceptibility curve at low temperatures. In order to see this explicitly we plot the magnetic susceptibility X_m for $\varepsilon_f=0$ as a function of the temperature in Fig.11 [31]. It is seen that the susceptibility curve has a maximum X_{max} at a finite temperature and the value of X_{max}/X_m (T=0) increases as J increases [31]. In the Coqblin-Schrieffer model $(-\varepsilon_f \gg \Delta)$, the susceptibility curve is universal and the value of X_{max}/X_m (T=0) is also universal for a fix value of J [28]. As ε_f approaches the Fermi level, the effect of the charge fluctuation becomes conspicuous and the susceptibility curves lead to those shown in Fig.11 (while X_m (T=0) decreases) [31].

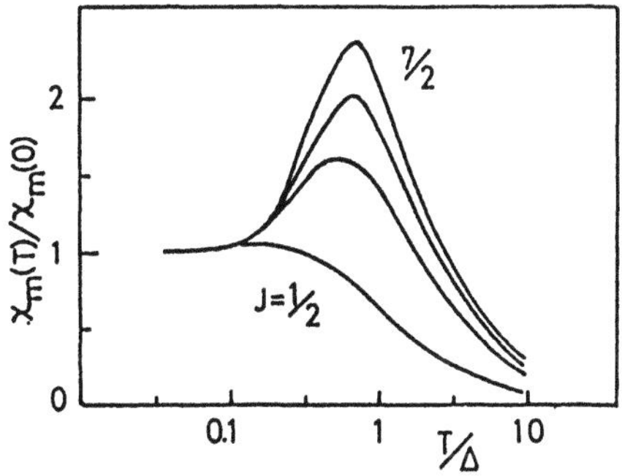

Fig.11. Plots of $X_m(T)/X_m(0)$ vs $\log(T/\Delta)$ for $\varepsilon_f=0$. The values of $\Delta X_m(0)$ for $J=1/2, 3/2, 5/2$ and $7/2$ are 0.17, 0.45, 0.89 and 1.5 , respectively.

3.3 Thermodynamic property of Ce impurity in a cubic crystalline field

Here, we describe the thermodynamic properties of Ce impurity in a cubic
crystalline field in detail. For a Ce atom, the large spin-orbit coupling
leads to the ground state with the total angular momentum $J=5/2$ and the
excited state of $J=7/2$ lying about 300 meV higher up. We neglect the
effect of the $J=7/2$ state, which is not expected to play an important role
at low temperatures although the effect of $J=7/2$ state is included easily
in a similar way to below. The $J=5/2$ ground multiplet splits into the
Γ_7-doublet and the Γ_8-quartet in a cubic crystalline field and these states
split further in the presence of the external magnetic field. In this case
the eigenvalue W_j of H_W is given by ($1 \leq j \leq 6$)

$$W_j = \begin{cases} \frac{1}{3}\Delta_c \pm \frac{1}{2}H \quad , \\[2ex] -\frac{1}{6}\Delta_c + \frac{1}{2}H \pm \{\frac{1}{4}\Delta_c^2 + \frac{4}{3}\Delta_c H + 4H^2\}^{\frac{1}{2}} \quad , \\[2ex] -\frac{1}{6}\Delta_c - \frac{1}{2}H \pm \{\frac{1}{4}\Delta_c^2 - \frac{4}{3}\Delta_c H + 4H^2\}^{\frac{1}{2}} \quad , \end{cases} \tag{3.10}$$

where Δ_c represents the energy splitting due to a crystalline field ($\Gamma_7(\Gamma_8)$
is ground state for $\Delta_c>0$ ($\Delta_c<0$)) and H is the magnetic field applied in the
z-direction ($g\mu_B=1$). Note that W_j should be assigned one of the six values
in (3.10) so as to satisfy the inequality (3.2).

With the use of (3.4), (3.5) and (3.10), we can evaluate the thermodynamic
quantities of the present system for arbitrary values of T, Δ_c and H [35,36].
Henceforth we are concerned with the case for $\Delta_c>0$, Γ_7 ground state. The
contribution from H_W to the thermodynamic potential is given by [35]

$$\begin{aligned} W &= \sum_{j=1}^{\nu} W_j(N_j-N_{j+1}) \\ &= \sum_{j=1}^{\nu} \{\sum_{i=1}^{j} W_i \int_{-\infty}^{\infty} \rho_0^{(j)}(k)dk + (W_j-W_{j-1}) \sum_{n=1}^{\infty} n \int_{-\infty}^{\infty} \rho_n^{(j)}(k)dk\} \quad . \end{aligned}$$

The expansion of W up to H^2 gives the expression for the magnetic
susceptibility χ_m at $T=0$ [34,35],

$$\chi_m = \chi_7 + \chi_8 + \chi_V \tag{3.11a}$$

$$\chi_7 = (25/36\pi)\lim_{Q_1 \to -\infty} \{\rho_I^{(1)}(Q_1)/\rho_C^{(1)}(Q_1)\} \quad , \tag{3.11b}$$

$$\chi_8 = (65/54\pi)\sum_{j=3}^{5} \lim_{Q_j \to -\infty} \{\rho_I^{(j)}(Q_j)/\rho_C^{(j)}(Q_j)\} \quad , \tag{3.11c}$$

$$\chi_V = (80/9\Delta_c)\int_{-\infty}^{Q_2} \rho_I^{(2)}(k)dk \quad , \tag{3.11d}$$

where χ_7, χ_8 and χ_V are the contributions to χ_m from the Γ_7 state, the Γ_8 state and the Van Vleck paramagnetism. The explicit expression for χ_m is given in ref.34. Furthermore the expression for the T-linear specific heat can be written as [35]

$$\gamma = \frac{6\pi^2}{25}\, \chi_7 + \frac{9\pi^2}{65}\, \chi_8 + \frac{\pi^2}{18}\, \chi_C + \frac{\pi^2}{6}\, \chi_\Delta \quad . \tag{3.12}$$

For Γ_8 ground state, a similar expression for the local Fermi-liquid relation is obtained in which χ_Δ is replaced by $2\chi_\Delta$ in (3.12). Here, χ_C and χ_Δ are the charge susceptibilies defined by $\chi_C = -\partial(n_7+n_8)/\partial\epsilon_f$ and $\chi_\Delta = \partial(n_7-n_8)/\partial\Delta_C$, where $n_7(n_8)$ is the occupation number of the Γ_7 state (Γ_8 state) [35]. Note that the Van Vleck susceptibility χ_V does not appear in this relation as a matter of course. In the absence of the crystalline field, χ_7, χ_8 and χ_Δ are in proportion to the magnetic susceptibility χ_m for $\Delta_C=0$, that is, $\chi_7=(5/63)\chi_m$, $\chi_8=(26/63)\chi_m$ and $\chi_\Delta=(4/35)\chi_m$. Therefore the relation (3.12) is reduced to the expression (3.9).

Hereafter we describe only the case of Coqblin-Schrieffer limit ($-\epsilon_f \gg \Delta$). The charge fluctuation of the localized electron is depressed in this limit, so that χ_C tends to zero in (3.12). In the absence of the crystalline field, this expression leads to

$$\gamma = (2\pi^2/21)\chi_m \quad , \quad \chi_m = 35/(4\pi T_0) \quad ,$$

where T_0 is the Kondo temperature (ν-fold degeneracy) obtained by

$$T_0 = \frac{1}{2\nu\Delta}\, \exp(\pi\epsilon_f/\nu\Delta)\int_{Q_\nu}^{\infty} \exp(-\pi k/\nu\Delta)\epsilon_0^{(\nu)}dk \quad . \tag{3.13}$$

Here, $\epsilon_0^{(\nu)}$ which represents the charge fluctuation, is not connected with other $\epsilon_0^{(j)}$ and is given by the following equation at $T=0$:

$$\epsilon_0^{(\nu)}(k) = k + \int_{Q_\nu}^{\infty} R(k-k')\epsilon_0^{(\nu)}(k')dk' \quad ,$$

where the cut-off parameter Q_ν is defined by $\epsilon_0^{(\nu)}(Q_\nu)=0$. Note that T_0 defined by (3.13) is related to the impurity contribution to the T-linear specific heat, C_i, through $C_i = \pi(\nu-1)T/6T_0$. In the absence of the magnetic field, the basic equations (3.7) and (3.8) can be solved with the aid of the standard Wiener-Hopf technique [26]. As a result, the coefficient of the T-linear specific heat in a closed form for Γ_7 ground state ($\Delta_C>0$) is

$$\gamma = \gamma_7 + \gamma_8 + \gamma_\Delta \tag{3.14a}$$

$$T_0\gamma_7 = \frac{1}{24}\left(\frac{3}{\pi^3}\right)^{\frac{1}{2}}\Gamma\left(\frac{1}{6}\right)^3\left(\frac{\Delta_C}{T_0}\right)^2\left[1+\left(\frac{T_0}{a\Delta_C}\right)^3 f\left(4,\frac{\pi i}{2}\right)\right] \quad , \tag{3.14b}$$

$$T_0\gamma_8 = \frac{3\pi^2}{8}\left(\frac{3}{2}\right)^{\frac{1}{2}}e^{-\frac{1}{4}}\Gamma\left(\frac{5}{4}\right)^{-1} f\left(4,\frac{\pi i}{4}\right)\frac{T_0}{\Delta_C} \quad , \tag{3.14c}$$

$$T_0\gamma_\Delta = \frac{1}{12a\sigma_2(0)}\frac{T_0}{\Delta_C}\int_{-\infty}^{\infty}\exp(-i0^+\omega)f(4,\omega)d\omega \quad , \quad \text{where} \tag{3.14d}$$

$$f(n,\omega) = \frac{1}{G(\omega)}\,\frac{1}{2\pi i}\int_{-\infty}^{\infty}\frac{dx}{x-\omega}\,G(x)\,\frac{\sinh(nx)}{\sinh(6x)}\left(\frac{a\Delta_c}{T_0}\right)^{-6ix/\pi}\ ,$$

$$G(\omega) = \frac{\sqrt{3}}{2}\left[3^6(-i\omega+0)/2^4e\pi\right]^{-\frac{i\omega}{\pi}}\Gamma(1-\frac{2i\omega}{\pi})\Gamma(1-\frac{4i\omega}{\pi})/\Gamma(1-\frac{i\omega}{\pi})\Gamma(1-\frac{6i\omega}{\pi})\ .$$

Here, $\Gamma(x)$ represents the gamma function, $a=(3e/2)^{1/6}\Gamma(1/6)/3\pi$ and $\sigma_2(k)=\rho_c^{(2)}(\Delta k+Q_2)/T_0$. The contribution to γ from the $\Gamma_7(\Gamma_8)$ state is written as $\gamma_7(\gamma_8)$ and γ_Δ is due to the excitation from Γ_7 state to the Γ_8 state. For Γ_8 ground state ($\Delta_c<0$) similar expressions for γ_7, γ_8 and γ_Δ are obtained, which are given in ref.35.

For the large crystalline field splitting, we define the effective Kondo temperature by $T_e=\pi(\nu_e-1)/6\gamma$, where the ground multiplet is assumed to be ν_e-fold degenerate. When $\Delta_c\gg T_0$ ($\nu_e=2$, Γ_7 ground state), we obtain from (3.14) the expression for T_e as [36]

$$T_e^{(7)}/T_0 = 4(\pi^5/3)^{\frac{1}{2}}\Gamma(1/6)^{-3}(\Delta_c/T_0)^{-2}\sim 0.234(\Delta_c/T_0)^{-2}\ ,\tag{3.15}$$

and when $-\Delta_c\gg T_0$ ($\nu_e=4$, Γ_8 ground state), we obtain

$$T_e^{(8)}/T_0 = 8\sqrt{\pi}\,6^{-\frac{1}{4}}\Gamma(5/4)\Gamma(1/6)^{-\frac{3}{2}}(|\Delta_c|/T_0)^{-\frac{1}{2}}\sim 0.624(|\Delta_c|/T_0)^{-\frac{1}{2}}\ .\tag{3.16}$$

These exact expressions for the Kondo temperature are useful even if the value of $|\Delta_c|/T_0$ is not so large. For the large crystalline field γ can be expressed by

$$\gamma = (6\pi^2/25)\chi_7\ ,\quad \chi_7 = 25/(36\pi T_e^{(7)})\ ,\quad \Delta_c\gg T_0\ ,$$

$$\gamma = (9\pi^2/65)\chi_8\ ,\quad \chi_8 = 65/(18\pi T_e^{(8)})\ ,\quad -\Delta_c\gg T_0\ .$$

The numerical results for γ are shown in Fig.12(a) as a function of Δ_c/T_0 [36]. It is seen that the value of γ is increased with increasing value of

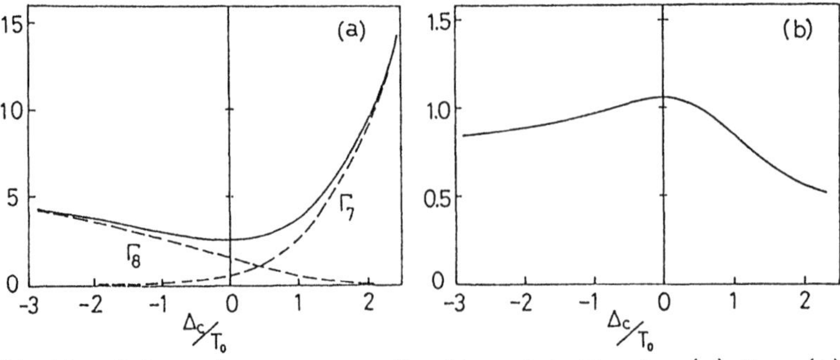

Fig.12. Universal curves as a function of Δ_c/T_0, for (a) $T_0\gamma$, (b) $R=\chi_m/\gamma$. Dashed curves in (a) represent the contribution for the Γ_7 (Γ_8) state.

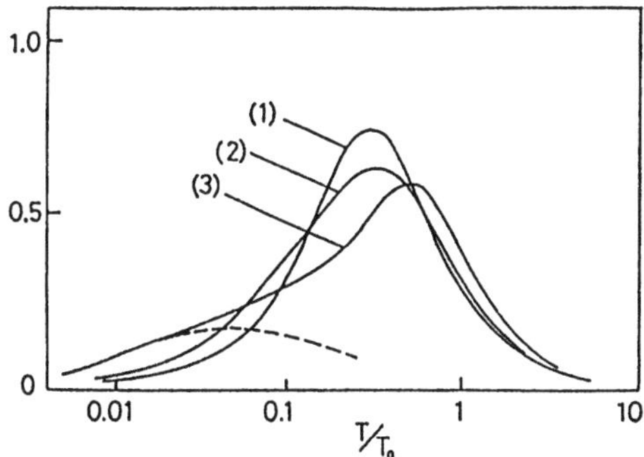

Fig.13. Plots of specific heat as a function of T/T_0 for the Γ_7 ground state: (1) $\Delta_C/T_0=0$, (2) 1, (3) 2. Dashed curve represents the contribution from Γ_7 state for $\Delta_C/T_0=2$.

$|\Delta_C|$ and is dominated by the contribution from $\Gamma_7(\Gamma_8)$ state for $\Delta_C>0(\Delta_C<0)$ even if $|\Delta_C|$ is not so large compared to T_0.   For the magnetic susceptibility, this is not true, because the large contribution from the Van Vleck susceptibility remains even for the rather large value of $|\Delta_C|/T_0$[34].

The results for Wilson's ratio $R=\chi_m/\gamma$ are shown in Fig.12(b). The value of R changes from $21/2\pi^2$ to its asymptotic value $25/6\pi^2$ $(65/9\pi^2)$ for $\Delta_C>0$ $(\Delta_C<0)$ with the increase of $|\Delta_C|$. This dimensionless quantity has been often used to investigate the experimental results at low temperatures [33]. It is noted that the rather small crystalline field splitting affects the value of R considerably, especially for $\Delta_C>0$.

The numerical calculation at finite temperatures is carried out by truncating the integral equations (3.4) at finite numbers n and converting the integral equations into a set of the matrix equations in which each pseudoenergy and its derivative are represented by a few hundred lattice points. These matrix equations are solved by the iteration method. The specific heat obtained in the Coqblin-Schrieffer limit is shown in Fig.13 as a function of the temperature for Γ_7 ground state $(\Delta_C>0)$[36]. It is seen that the shape of the specific heat curve becomes asymmetric and has a shoulder structure at the low temperature side, when the value of Δ_C is gradually increased. This shoulder structure is caused by the low-lying excitation states within the Γ_7-doublet. The contribution to C_i from the Γ_7-doublet is also shown for $\Delta_C/T_0=2$ in Fig.13. When the value of Δ_C is further increased, the specific heat curve may have two peaks, as would be expected. In that case, the pattern behaviour at high temperatures is described approximately by Schottky formula

$$C_i = 2(\Delta_C/T)^2\exp(\Delta_C/T)[\exp(\Delta_C/T)+2]^{-2} \quad .$$

On the other hand, the specific heat at low temperatures shows the singlet bound state nature.   They are described by the s-d exchange model (spin 1/2), in which the effect of the crystalline field does not appear explicitly but only through the effective Kondo temperature $T_e^{(7)}$ defined by (3.15). The results for Γ_8 ground state are given in ref.36. The shoulder structure similar to the curve 3 in Fig.13 has been found experimentally for $CeCu_{2.2}Si_2$ and $CeA\ell_3$ [44] (see Fig.14). Although the crystalline field is not cubic but is tetragonal or hexagonal in these materials, the essential properties of the specific heat might be explained with the results mentioned here, that is, in both materials the shoulder structure is caused by the low-lying excitation within the ground-doublet and the peak structure is caused by the direct excitation from the ground doublet to the excited multiplets.

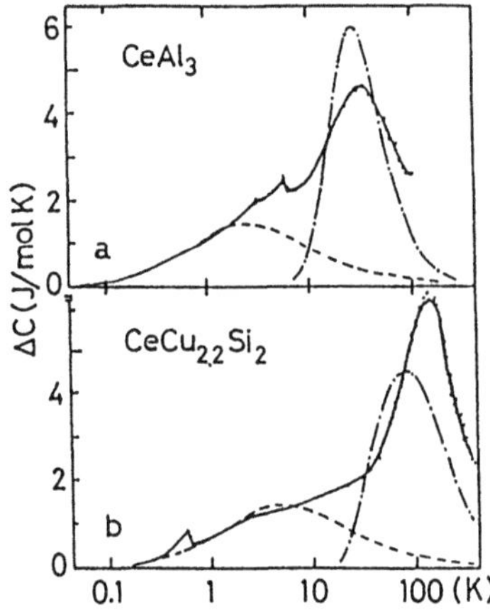

Fig.14. 4f-derived specific heat ΔC vs temperature on a logarithmic scale for polycrystalline samples of CeAℓ_3 (a) and CeCu$_{2.2}$Si$_2$ (b). Theoretical ΔC results are included for comparison: (i) Bethe Ansatz for spin $\frac{1}{2}$ (broken line); (ii) Schottky anomalies (dash-dotted line) calculated for three crystal-field doublets at OK, 57K, 84K (CeAℓ_3), and at OK, 140K, 364K (CeCu$_{2.2}$Si$_2$). Note that the anomalies in C($\bar{T}$) of CeAℓ_3 near 3.5K and 6K must be ascribed to the onset of magnetic order in spurious precipitations of CeAℓ_2 and Ce$_3$Aℓ_{11} (after F. Steglich et al., ref.44).

Finally we show some calculated results of the magnetic properties for the Coqblin-Schrieffer model at zero temperature for reference. The magnetic properties in this case can also be calculated in a similar method employed in the single-orbital case. The magnetization curve M(H) is shown for the case of Γ_7 ground state in Fig.15 [45]. In the presence of the crystalline field (curve 3 in Fig.15) M(H) has a steep rise structure at a finite magnetic field [38,45,46]. This structure is caused by the crossing of two low-lying energy levels, in which the effective moment is different from each other.   For the magnetoresistance, on the other hand, the effect of level crossing results in the peak structure at the same value of the applied magnetic field [38,46]. The origin of the peak structure appearing in the magnetoresistance is considered to be that the strong magnetic field produces the sharp peak of the state density at the Fermi level, when the level-crossing occurs. In fact, the formation of the sharp peak at the Fermi-level is shown directly with the calculated results of the magnetic field-dependent specific heat coefficient (see Fig.16)[45].

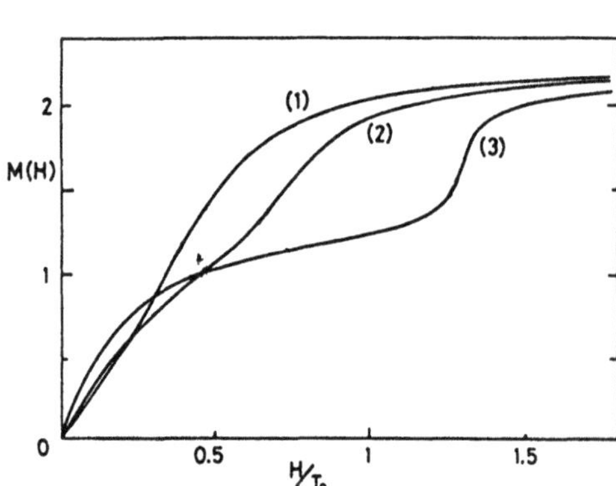

Fig.15. Magnetization M(H) for the Γ_7 ground state: (1) Δ_c/T_0=0, (2) 1, (3) 2.

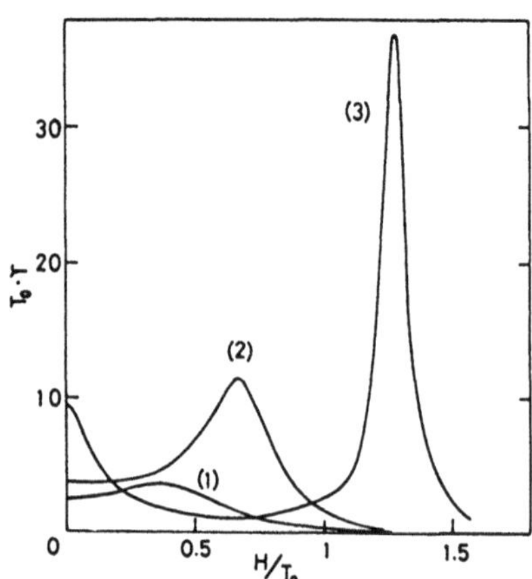

Fig.16. Coefficient of T-linear specific heat γ for the Γ_7 ground state: (1) Δ_c/T_0=0 , (2) 1, (3) 2.

4. The transport and the dynamical properties of the Anderson model

4.1. Introduction

The third part of this article is about the transport and the dynamical
properties of the Anderson model. In general it is impossible to calculate
the physical quantities related to the transport phenomena and dynamics
with the use of the Bethe Ansatz solution. Nevertheless, using the local
Fermi-liquid theory developed for the impurity problem [7,13,40,48-50],
we can calculate some of them at low temperatures and low frequencies
precisely [51]. The basic idea used for the calculation of the transport
coefficients is the conventional one, that is, Friedel sum-rule and the
Fermi-liquid concept for the impurity system which can relate the transport
coefficients to the static quantities at low temperatures.

The magnetoresistance for the single orbital Anderson model is given by
the well-known formula

$$R(H)^{-1} = (2R_0)^{-1} \sum_\sigma \frac{1}{\pi \Delta \rho_{d\sigma}(0)} \quad ,$$

where R_0 is the value for $\varepsilon_d = U = 0$ and $\rho_{d\sigma}(0)$ is the impurity part of the
density of states at the Fermi level. The magnetic field dependent Hall
coefficient is also expressed in terms of $\rho_{d\sigma}(0)$,

$$R_H(H)/R_H(0) = 2\sum_\sigma \rho_{d\sigma}^{-2}(0)/(\sum_\sigma \rho_{d\sigma}^{-1}(0))^2 \quad .$$

In the derivation of the above expressions we have used the fact that the
relaxation time $\tau_\sigma(\varepsilon)$ for each spin is inversely proportional to $\rho_{d\sigma}(\varepsilon)$[52].
The numerical results of the magnetoresistance and the Hall coefficient as
a function of the external magnetic field for the single orbital Anderson
model of $U \to \infty$ are shown in Figs.17 and 18 [51]. Here, $\tilde{\varepsilon}_d$ is the effective
impurity level introduced by Haldane [53]. In the empty-orbital regime
$\tilde{\varepsilon}_d \gg \Delta$, $R(H)$ shows similar properties as in the $U=0$ case, because the
correlation effect is not so important in this regime. The correlation
effect becomes conspicuous if the effective d-level position is lowered
through the Fermi level. A sharp drop of $R(H)$ appears at small magnetic
field [49], which can be expressed in the Kondo regime by

$$R(H) - R(0) \sim -(\tilde{\chi}/\Delta)^2 H^2 \quad ,$$

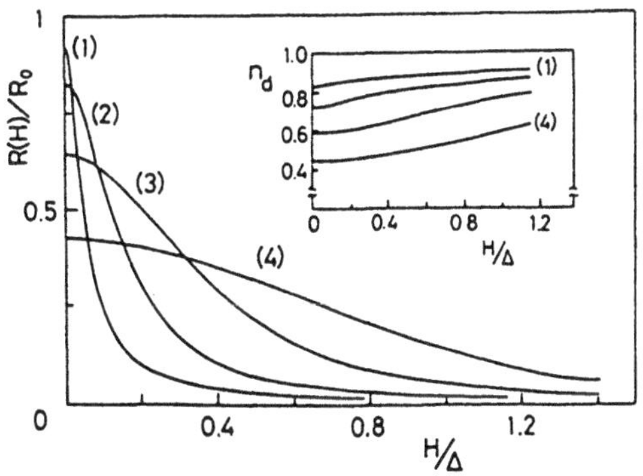

Fig.17. Plots of magnetoresistance:
(1) $\tilde{\varepsilon}_d/\Delta = -1.6$, (2) -0.8, (3) 0,
(4) 0.8. The field dependence of
the localized electron number is
shown in the inset.

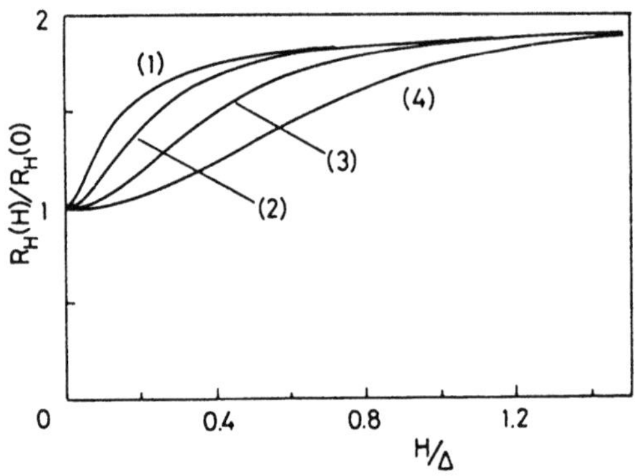

Fig.18. Plots of field-dependent Hall coefficient: (1) $\tilde{\varepsilon}_d/\Delta=-1.6$, (2) -0.8, (3) 0, (4) 0.8.

where $\tilde{\chi}$ is the enhancement factor written by Δ/T_K (T_K is the Kondo temperature) [49]. The value of $R_H(H)/R_H(0)$ is seen to change from 1 to 2. If the relaxation time $\tau_\sigma(0)$ takes the same value for both spins, $R_H(H)/R_H(0)$ equals unity. As the magnetic field is increased, the value of $\tau_\sigma(0)$ for one spin direction becomes large with respect to that for the other, so that $R_H(H)/R_H(0)$ approaches two. The sharp-rise structure at weak fields for $\tilde{\varepsilon}_d<0$ is caused by the large enhancement factor. The field dependence of the density of states $\rho_{d\sigma}(0)$ for each spin direction is shown in Fig.19, which has been used for the evaluation of the magnetoresistance and the Hall coefficient [51]. It is seen that the difference of the state density for each spin-direction is large at small magnetic fields in the Kondo regime.

Finally, we mention that the thermal conductivity κ can also be expressed in terms of $\rho_d(0)$ as

$$\kappa = \kappa_0 \, T/\rho_d(0) \quad ,$$

where κ_0 is a constant. Then the κ is written as $\kappa=\alpha T/R(0)$ (Wiedemann-Franz law), where α is the Lorentz number. The field dependence of κ is proportional to $R^{-1}(H)$.

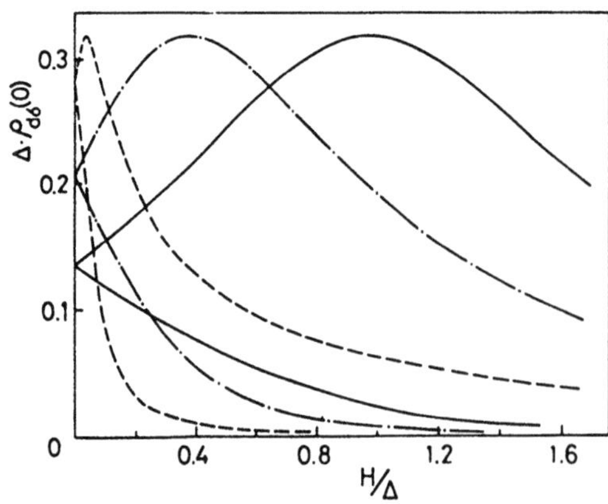

Fig.19. Plots of $\rho_{d\sigma}(0)$ for each spin as a function of H/Δ: $\tilde{\varepsilon}_d/\Delta=-1.2$ (broken line), 0 (dashed-dotted line), 0.8 (solid line).

4.2 Thermoelectric power at low temperature [51,54]

First we consider the thermoelectric power of the single orbital Anderson model with finite U. According to the Boltzmann equation, the thermoelectric power (TEP) is described in terms of the Fermi function $f(\varepsilon)$ and the relaxation time $\tau(\varepsilon)$ of the conduction electron [51],

$$S(T) = - \frac{1}{eT} \int_{-\infty}^{\infty} \frac{\partial f}{\partial \varepsilon} \, \varepsilon\tau(\varepsilon)d\varepsilon \Big/ \int_{-\infty}^{\infty} \frac{\partial f}{\partial \varepsilon} \, \tau(\varepsilon)d\varepsilon \quad , \tag{4.1}$$

resulting in the expression for the linear term with respect to the temperature $T(k_B=1)$

$$S = -(S_0/2\pi)T\tau'(0)/\tau(0) \quad , \tag{4.2}$$

where $S_0=2\pi^3/3e$, prime means the energy-derivative and the energy is measured from the Fermi level. For the Anderson model, since the relaxation time $\tau(\varepsilon)$ is inversely proportional to the density of states $\rho_d(\varepsilon)$, the expression (4.2) can be rewritten as

$$S = (S_0/2\pi)T\rho_d'(0)/\rho_d(0) \quad . \tag{4.3}$$

Here, $\rho_d'(0)$ means the energy derivative of $\rho_d(0)$ at the Fermi level and is expressed as

$$\rho_d'(0) = \frac{2\pi}{\Delta} \, \tilde{\chi}(\varepsilon_d+\Sigma_d(0)) \rho_d(0)^2 \quad ,$$

where $\tilde{\chi}$ is the enhancement factor given by [48]

$$\tilde{\chi} = 1-\partial\Sigma_d(\varepsilon)/\partial\varepsilon\Big|_{\varepsilon=0} = \tilde{\gamma}/\rho_d(0)$$

and $\Sigma_d(\varepsilon)$ is the self-energy for the d-electron which is expressed by [4,7, 48-50]

$$\Sigma_d(0) = -\varepsilon_d+\Delta\cot\delta_\sigma \quad , \qquad \rho_d(0) = (\pi\Delta)^{-1}\sin^2\delta_\sigma \quad ,$$

and $\tilde{\gamma}$ means the specific-heat coefficient normalized at $U=\varepsilon_d=0$ and δ_σ is the phase shift at the Fermi level. Then we have [51,54]

$$S = S_0\tilde{\gamma}T \cot\delta_\sigma \quad . \tag{4.4}$$

As a result, the exact expressions necessary for the calculation of TEP are [5,6]

$$\tilde{\gamma} = \frac{1}{4\pi} \left[\frac{\sigma_I(a)}{\sigma_C(a)} + \frac{\chi_s +\int_{-\infty}^{a} e^{\pi x}\sigma_I(x)dx}{1/2\pi+\int_{-\infty}^{a} e^{\pi x}\sigma_C(x)dx} \right]$$

$$n_d = 1 - \int_{-\infty}^{a} \sigma_I(x)dx \quad ,$$

where χ_s is the magnetic susceptibility for the symmetric case and $\sigma_I(\sigma_C)$ is the distribution function of the impurity part (conduction-electron part) at $T=0$, which is given by the solution of the linear integral equation as has

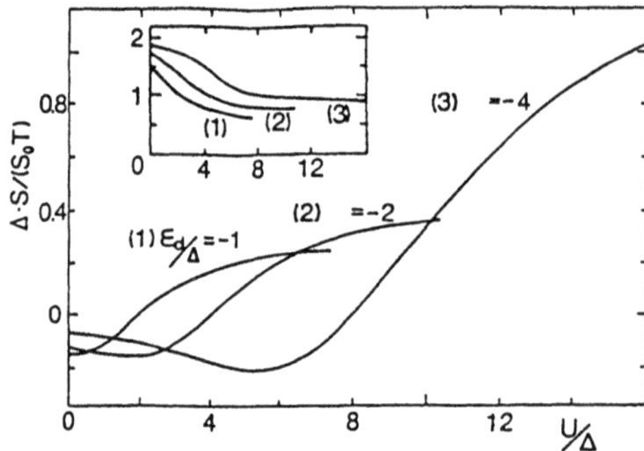

Fig.20. Plots of T-linear coefficient of TEP as a function of U/Δ. The d-electron number n_d is shown in the inset.

been mentioned in chapter 2. Note that a, which corresponds to the Fermi level for the charge excitation, is determined from the minimization condition for the free energy [5,6].

The numerical results for TEP are plotted in Fig.20 as a function of U for several values of ε_d [54]. In order to see the characteristic properties of TEP more clearly we show in Fig.21 the results for the enhancement factor $\tilde{\chi}$, the density of states $\rho_d(0)$ and the renormalized d-level ε_d^* defined by

$$\varepsilon_d^* = (\Delta/\tilde{\chi})\cot\delta_\sigma \quad . \tag{4.5}$$

For the case of small U, the correlation effect is not so important that the enhancement does not occur ($\tilde{\chi}\sim 1$) and the renormalized d-level corresponds to the bare d-level ($\varepsilon_d^*\sim\varepsilon_d$). In this case, the correction due to the Coulomb interaction to TEP can be described well by the lower order perturbation expansion

$$S/T = S(U=0)/T + \frac{S_0(\Delta^2-\varepsilon_d^2)}{2\pi(\Delta^2+\varepsilon_d^2)^2}\, n_d(U=0)U + \cdots \quad ,$$

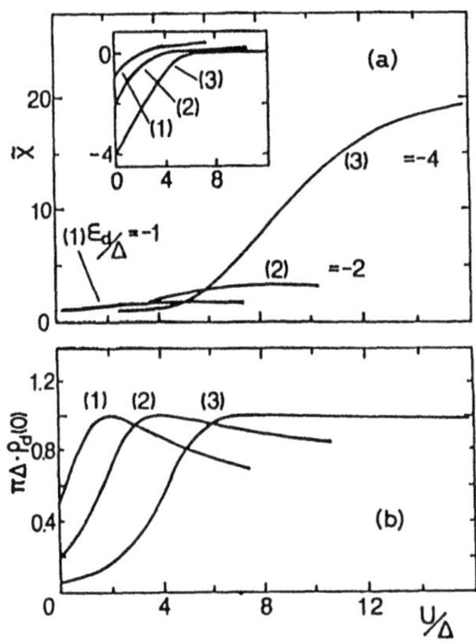

Fig.21. (a) Enhancement factor $\tilde{\chi}$, (b) density of states $\rho_d(0)$ as a function of U/Δ. The renormalized d-level ε_d^*/Δ is shown in the inset.

where TEP for the noninteracting case is

$$S(U=0)/T = (S_0/\pi)\varepsilon_d(\varepsilon_d^2+\Delta^2)^{-1} \quad .$$

As the Coulomb interaction is increased, the situations become somewhat
different and the values of S/T depend on the position of ε_d considerably.
In the case of $|\varepsilon_d|<\Delta$ (curve (1), so-called mixed valence regime), the
enhancement factor does not change so much even if the Coulomb interaction
becomes large. This is because the correlation effect in this regime
results in pulling up the renormalized d-level above the Fermi level rather
than increasing the $\tilde{\chi}$ value [53]. Hence, TEP is not so enhanced in this
case. On the other hand, when the bare d-level is sufficiently below the
Fermi level (curve (3)), the enhancement factor becomes very large with
increasing U, leading to the large positive value of TEP. This enhancement
is interpreted as a consequence of the development of the Kondo resonance
at the Fermi level. It is seen from the curve (3) in Fig.21 that the
renormalized d-level is fixed to the Fermi level and the density of states
approaches $1/\pi\Delta$. Note that the renormalized resonance width Δ^* is given by
$\Delta/\tilde{\chi}$, the value of which becomes order of T_K. The factor relevant to the
sign of TEP is $\cot\delta_\sigma$ in (4.4) or the renormalized d-level position ε_d^*
defined by (4.5), which reflects the asymmetry of $\rho_d(\varepsilon)$ near the Fermi level.
Since the electron-hole symmetry holds for the symmetric case ($U=-2\varepsilon_d$), the
value of TEP is equal to zero ($\varepsilon_d^*=0$). The deviation from the symmetric case
leads to either positive or negative TEP: S>0 for $U/2+\varepsilon_d>0$ and S<0 for
$(U/2)+\varepsilon_d<0$. If we denote the amount of the deviation from the symmetric
case as $\delta E=((U/2)+\varepsilon_d)/\Delta$, we can expand S with respect to δE with the aid of
the Wiener-Hopf technique. The resulting expression for S is

$$S/T = (S_0/\Delta)\tilde{\chi}\tilde{\chi}_C\delta E+O(\delta E^2) \quad , \quad \text{where}$$

$$\tilde{\chi}_C = \frac{1}{\pi}\int_{-\infty}^{\infty}\frac{\Delta/2U}{\left(y+\sqrt{U/8\Delta}\right)^2+\Delta/2U}\ e^{-\pi y^2}dy \quad .$$

In the case of large U (s-d limit), $\tilde{\chi}\tilde{\chi}_C$ takes a very large value because
in this parameter regime $\tilde{\chi}$ and $\tilde{\chi}_C$ are given by [5,6]

$$\tilde{\chi} \sim \pi(\Delta/2U)^{\frac{1}{2}} \exp(\pi U/8\Delta) \quad ,$$

$$\tilde{\chi}_C \sim (4/\pi)(\Delta/U)^2 \quad .$$

Therefore a little asymmetry of $\rho_d(\varepsilon)$ of the system leads to the giant TEP
with either positive or negative sign. It is considered that the ordinary
potential scattering easily gives such a relaxation-time asymmetry [55].

Next, we investigate the T-linear coefficient of TEP for the highly
correlated degenerate Anderson model with the angular momentum J and also
the case including the cubic crystalline field [54,56]. In that case the
enhancement factor can be written as

$$\tilde{\chi}_m = \tilde{\gamma}/(\nu\rho_f(0)) \quad , \quad \text{for } \nu(=2J+1)\text{-fold degenerate case}$$

and $\quad \tilde{\chi}_7 = (\chi_7/2+\chi_\Delta/3)/\rho_7(0) \quad ,$

$$\tilde{\chi}_8 = (3\chi_8/4+\chi_\Delta/12)/\rho_8(0) \quad , \quad \text{for J=5/2 in the cubic field. Here,}$$

$\rho_f(0)$ means the density of states for the f-electron at the Fermi level and

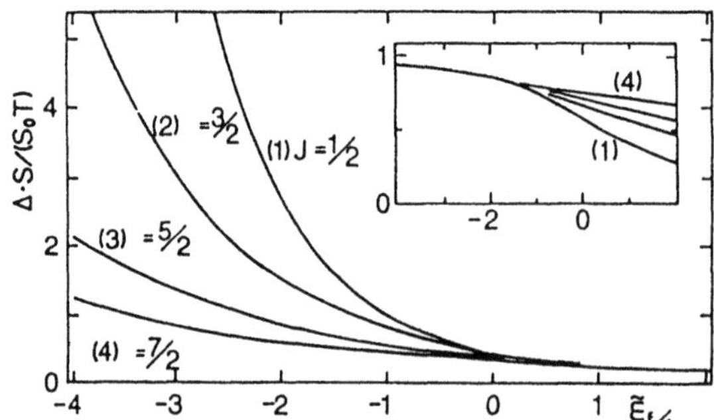

Fig.22. Plots of T-linear coefficient of TEP for the degenerate case. The f-electron number n_f is shown in the inset.

$\tilde{\gamma}$ is the specific heat coefficient. Further, $\chi_7(\chi_8)$ and $\rho_7(0)(\rho_8(0))$ are the spin susceptibility and the density of states at the Fermi level for the $\Gamma_7(\Gamma_8)$ state, respectively, and χ_Δ is the charge susceptibility between the Γ_7 and Γ_8 states (see Chapter 3).

First, we discuss the T-linear coefficient of TEP for highly correlated degenerate Anderson model without the crystalline field. For this model, the effective f-level introduced by Haldane [53], $\tilde{\varepsilon}_f$, is an appropriate parameter for the specification of the valence state, as has been stated before. In Fig.22, the T-linear coefficient of TEP calculated is shown as a function of $\tilde{\varepsilon}_f$ for the case of J=1/2, 3/2,$\cdots$, 7/2. Related quantities are shown in Fig.23, such as the enhancement factor, the density of states and the renormalized f-level. It is seen that the value of $\tilde{\chi}$ becomes very large in the Kondo regime ($n_f{\sim}1$, $\pi\Delta\rho_f(0){\sim}\sin^2[\pi/(2J+1)]$), reflecting the strong correlation effect. Hence TEP in this regime takes a large value, as has been discussed before. For the same value of $\tilde{\varepsilon}_f$, it is seen that TEP takes a larger value with the decrease of the degeneracy.

This fact arises from the degeneracy dependence of $\tilde{\chi}$, which is written as [6,37,57]

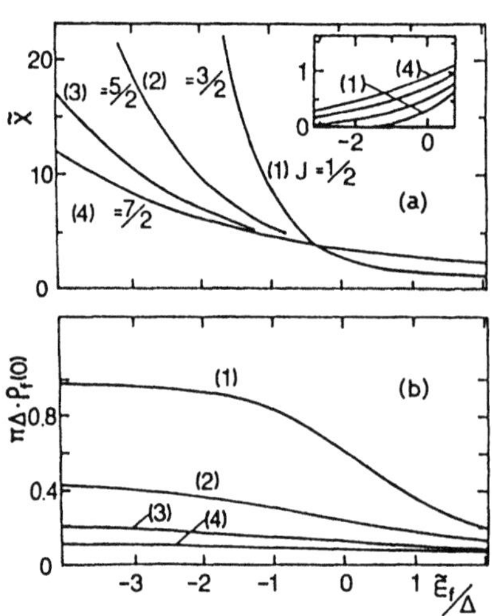

Fig.23. (a) Enhancement factor $\tilde{\chi}$, (b) density of states $\rho_f(0)$ for the degenerate case. The renormalized f-level ε_f^*/Δ is shown in the inset.

90

$$\tilde{\chi} \sim \exp[-\pi\tilde{\epsilon}_f/(2J+1)\Delta]$$

in the Kondo regime. If we define the Kondo temperature $T_K^{(J)}$ by $\Delta/\tilde{\chi}$, TEP is written as [54]

$$S/T = \frac{S_0}{\pi T_K^{(J)}} \cot(\pi n_f/(2J+1)) \quad ,$$

where $n_f \sim 1$. As discussed previously, the sign of TEP is determined by that of the renormalized f-level ϵ_f^*. The position of ϵ_f^* in the Kondo regime lies very near and always above the Fermi level as seen in Fig.23. For the case $J>1/2$, ϵ_f^* and the renormalized width Δ^* is given by [54]

$$\epsilon_f^* \sim T_K^{(J)} \cot(\pi/2J+1)$$
$$\Delta^* \sim T_K^{(J)} \quad .$$

Therefore it can be said that the sign of TEP is positive and is rather stable for $J>1/2$. The peak position ϵ_f^* of the Kondo resonance for $J=1/2$ may be much smaller than its width $T_K^{(\frac{1}{2})}$ due to the $\cot(\pi n_f/2)$ factor in the above expression, because $n_f \sim 1$ in the Kondo regime. Therefore the sign of TEP may be easily changed by the appearance of a small perturbation such as the potential scattering term. This situation is not changed in the case where the large cubic crystalline field is included and the ground state is a doublet of $J=5/2$ (see the inset in Fig.23). The T-linear coefficient of TEP for $J=5/2$ case with the cubic crystalline field is shown in Fig.24 [56], where Δ_C is the energy splitting by the cubic crystalline field and T_0 is the Kondo temperature defined at $\Delta_C=0$. It has been found experimentally that the sign of TEP for the heavy-electron Ce compounds shows rather complicated behaviours for its temperature dependence, in contrast to the Yb systems [58]. In these Ce systems the crystalline field ground state is the Kramers doublet with rather large crystalline field splitting compared

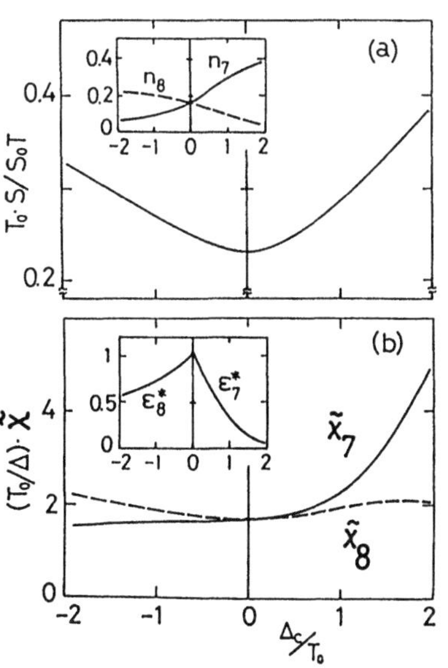

Fig.24. (a) T-linear coefficient of TEP as a function of the crystal-field splitting Δ_C/T_0, where T_0 is the Kondo temperature for the six-fold degeneracy and $S_0=2\pi^2/3e$. The occupation probabilities per channel n_7 and n_8 are shown in the inset.
(b) Specific-heat enhancement factors $\tilde{\chi}_7$ and $\tilde{\chi}_8$. The renormalized f-level ϵ_m^*/T_0 is shown in the inset.

to the Kondo temperature [58]. Therefore complicated behaviours of the sign
of TEP, especially at low temperature side, may be considered to come from
the fact that the peak position of the Kondo resonance ε_m^* lies very close to
the Fermi level [54,56].

4.3 The low-frequency dynamical susceptibility

The calculation for the low frequency dynamical susceptibility at zero
temperature in the crystalline field is carried out precisely with the use
of the Bethe Ansatz solution in order to investigate the dynamical properties
of the heavy-electron Ce compounds [59]. Following Shiba [60], the low
frequency susceptibility is obtained on the basis of the local Fermi liquid
theory as follows ($g\mu_B=1$):

$$\mathrm{Im}\ \chi^{+-}(\omega)/\omega = \pi \sum_{m\neq m'} (J_+)^2_{mm'}\rho_{fm}(0)\rho_{fm'}(0)[K^{mm'}]^2 \quad ,$$

$$\mathrm{Im}\ \chi^{zz}(\omega)/\omega = \pi \sum_{mm'} (J_z)^2_{mm'}\rho_{fm}(0)\rho_{fm'}(0)[K^{mm'}]^2 \quad ,$$

where $\rho_{fm}(0)$ is the density of states of the f-electron at the Fermi level,
$(J_+)_{mm'}$ and $(J_z)_{mm'}$ are matrix elements of J_+ and J_z with the total angular
momentum J. It is easy to extend Shiba's calculation to the case where the
crystalline field and the magnetic field are included [59]. In that case,
it is necessary to calculate the vertex part $K^{mm'}$ for $\varepsilon_{fm}\neq\varepsilon_{fm'}$. The obtained
expression for $K_{mm'}$ which we call the generalized enhancement factor is[40,47]

$$K^{mm'} = \begin{cases} 1 + \dfrac{\Sigma_m(0)-\Sigma_{m'}(0)}{\varepsilon_{fm}-\varepsilon_{fm'}} & (\text{for } \varepsilon_{fm}\neq\varepsilon_{fm'}) \\[2em] \tilde{\chi}_m & (\text{for } \varepsilon_{fm}=\varepsilon_{fm'}) \end{cases} \quad ,$$

where $\Sigma_m(0)$ is the self-energy. The details for the derivation of these
expressions are given in ref.59.

The results for J=5/2 case in the presence of the cubic crystalline field
as a function of Δ_c/T_0 are shown in Fig.25. It is seen that the dynamical
susceptibility is enhanced as the value of the crystalline field is increased.
Further, in the strong field regime the main contribution to the dynamical
susceptibility comes from the low-lying multiplet, then we have

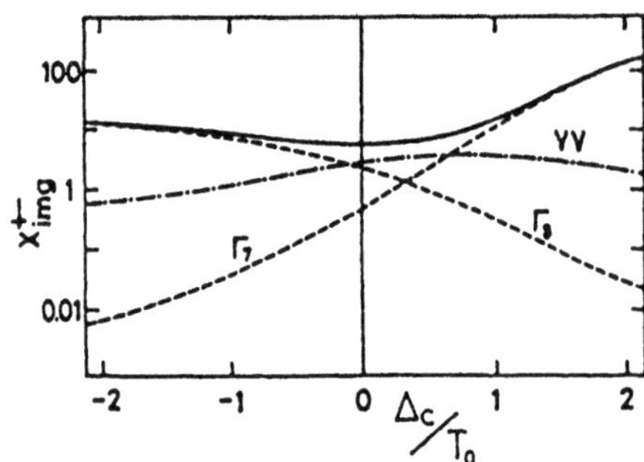

Fig.25. ω-linear term of the dynamical
susceptibility for the cubic field
(solid line): $\chi^+_{img}=T_0^2\mathrm{Im}[\chi^{+-}(\omega)]/[\omega(g\mu_B)^2]$.
The broken lines are the contribution
from the Γ_7 and Γ_8 states, and dash
line from the Van Vleck process (VV).
The quantity T_0 is the Kondo temperature
for $\Delta_c=0$.

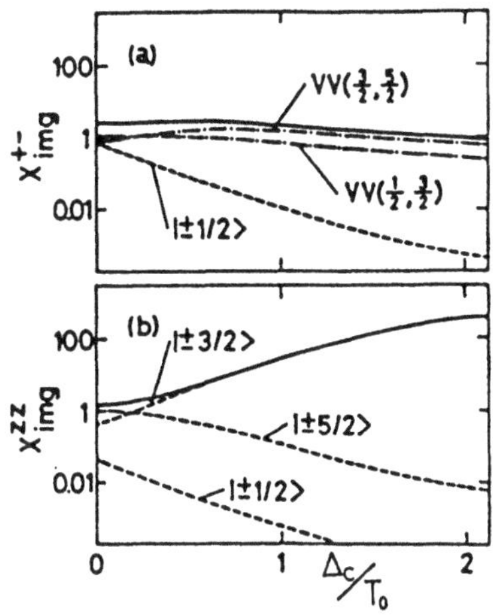

Fig.26. Plots of (a) $\chi^{+-}_{img}=T_0^2 Im[\chi^{+-}(\omega)]/[\omega(g\mu_B)^2]$ and (b) $\chi^{zz}_{img}=T_0^2 Im[\chi^{zz}(\omega)]/[\omega(g\mu_B)^2]$ for the hexagonal field (solid line). The ground doublet is $|\pm3/2\rangle$ and excited states $|\pm5/2\rangle$ and $|\pm1/2\rangle$ are separated from $|\pm3/2\rangle$ by Δ_c and $2\Delta_c$. The broken lines are the contributions from each doublet state and dash-dotted lines from the Van Vleck processes: VV(1/2,3/2) represents the Van Vleck term between $|\pm1/2\rangle$ and $|\pm3/2\rangle$.

$$
\lim_{\omega\to 0} \frac{Im[\chi^{+-}(\omega)]}{\omega} = \begin{cases} \dfrac{25}{36\pi}\left(\dfrac{1}{T_e^{(7)}}\right)^2 & \text{(for } \Gamma_7 \text{ ground state)} \\[4mm] \dfrac{65}{18\pi}\left(\dfrac{1}{T_e^{(8)}}\right)^2 & \text{(for } \Gamma_8 \text{ ground state)} \end{cases} ,
$$

where $T_e^{(7)}(T_e^{(8)})$ is the effective Kondo temperature given by (3.15)((3.16)). The isotropic relation ($\chi^{+-}(\omega)=2\chi^{zz}(\omega)$) holds in the cubic symmetry as a matter of course.

The numerical results in the case of the hexagonal crystalline field with the assumed level scheme is given in Fig.26. In the hexagonal crystalline field, there exists an anisotropy of the magnetic relaxation due to the localized f-electrons, which comes chiefly from the difference of the matrix element between J_+ and J_z within the ground doublet. For the case of this level scheme J_+ has no matrix element within the $J_z=\pm3/2$ doublet. That is, $\chi^{+-}(\omega)$ is dominated by the Van Vleck term while $\chi^{zz}(\omega)$ has the contribution from the ground doublet. It is seen that the value $\chi^{zz}(\omega)$ is enhanced remarkably while that of $\chi^{+-}(\omega)$ is decreased in the range $\Delta_c/T_0>1$, leading to the extremely large anisotropy [59]. These results may be applicable to explain the experimental findings, such as the quasielastic line-width of the neutron scattering and the relaxation time of NMR [61,62]. In this section we have investigated the low-frequency dynamical susceptibility for the Ce-Kondo system at zero temperature in the presence of the crystalline field. One of the characteristic properties due to the crystalline field is the enhancement of the low-frequency dynamical susceptibility. As another aspect of the crystalline field effects, there exists the anisotropy of the low-frequency susceptibility which depends considerably on the wave function of the ground multiplet [59].

4.4 The magnetic field dependence of the friction coefficient of an adatom on the metal surfaces

The final part of this chapter is about the magnetic field dependence of the friction coefficient of an adatom on the metal surfaces. In the dynamical process on the solid surfaces such as adsorption, desorption and diffusion, a crucial role is played by the dissipative force. For instance, in the process of adsorption, the moving atom loses its kinetic energy, because it suffers the dissipative force. At low temperatures the dissipative force for the adatom on the metal surfaces results mainly from the electronic fluctuation. In that case one can reduce the dissipative force to the friction force [63,64].

Here, we use the single orbital Anderson model in order to calculate the friction coefficient of adatoms on the metal surfaces. The friction coefficient η is given in terms of the force correlation function as follows [63]:

$$\eta = \frac{1}{Mk_B T} \ \mathrm{Re} \int_0^\infty <\hat{F} \ \hat{F}(t)> dt \quad ,$$

where M is the mass of the adatom, T the temperature and $<\cdots>$ denotes the thermal average. Here, for simplicity one-dimensional system is assumed for the motion of the adatom. Following Bohnen et al.[65], we may take F as

$$F = \sum_{k,\sigma} W_k (C^+_{k\sigma} C_{d\sigma} + C^+_{d\sigma} C_{k\sigma}) \quad , \quad \text{and}$$

$$W_k = -\partial V_k / \partial X \quad ,$$

where X is the coordinate of the adatom. In the presence of the magnetic field, η can be written down explicitly as [66]

$$\eta = \eta_1 + \eta_2 \quad ,$$

$$\eta_1 = \frac{2}{\pi M} \{\xi(0)/\Delta(0) - [\zeta(0)/\Delta(0)]^2\} \sum_\sigma \sin^2 \delta_\sigma \quad ,$$

$$\eta_2 = \frac{2}{\pi M} \frac{\zeta(0)}{\Delta(0)} \sum_\sigma (\Delta d\delta_\sigma / d\Delta)^2 \quad ,$$

where $\Delta(0)=\pi V^2 \rho(0)$, $\xi(0)=\pi W^2 \rho(0)$, $\zeta(\varepsilon)=\pi V W \rho(0)$ and $\rho(\varepsilon)$ means the density of states for the conduction electrons [67].

The numerical results obtained are shown in Fig.27 [67] for the symmetric Anderson model, where η_1 and η_2 are proportional to the dotted line and solid line respectively. One can see that η_1 decreases monotonically with increasing magnetic field as the magnetoresistance does in the dilute alloy as has been expected by the expression. On the other hand, η_2 shows the strong magnetic field dependence and has the peak structure. Further it is to be noted that the peak value of η_2 is ten times larger than that of η_1 in this parameter. Besides, the prefactor in the expression for η_2 is expected to be much larger than that of η_1. That is, we can conclude that the pattern behaviour of η as a function of the magnetic field is very similar to the solid line of the Fig.27, which shows the strong magnetic field dependence. From the results obtained in this calculation, it is suggested that the temperature dependence of the friction coefficient also shows very similar pattern behaviour. Finally, the asymmetric case is shown in Fig.28 where we choose the parameters of hydrogen atom on the transition metal surface. In this case, it is also expected that the friction coefficient shows the

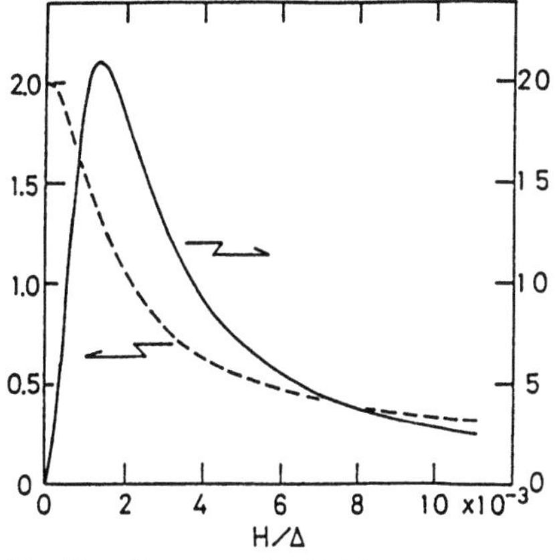

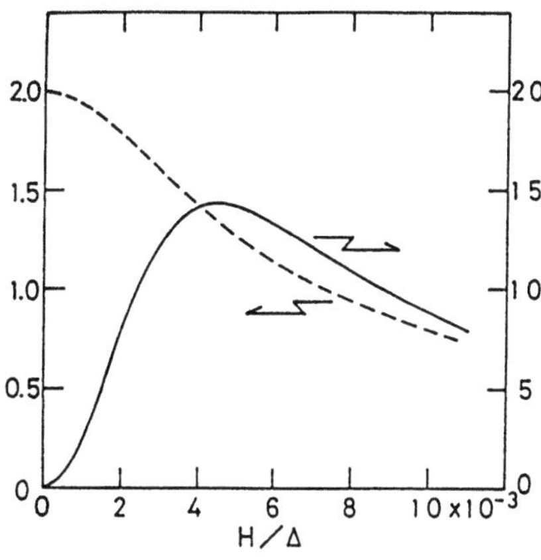

Fig.27. The magnetic field dependences of $\sum \sin^2\delta_\sigma$ (---) and $\sum(\Delta d\delta_\sigma/d\Delta)^2$ (——) for the symmetric case of $\varepsilon_d/2\Delta=-5$ and $U/2\Delta=10$. The maximum value of $\sum \sin^2\delta_\sigma$ is 2 at $H/\Delta=0$ and that of $\sum(\Delta d\delta_\sigma/d\Delta)^2$ is 21 at $H/\Delta=1.3\times10^{-3}$ in these parameters.

Fig.28. The magnetic field dependences of $\sum \sin^2\delta_\sigma$ (---) and $\sum(\Delta d\delta_\sigma/d\Delta)^2$ (——) for the asymmetric case of $\varepsilon_d/2\Delta=-7$ and $U/2\Delta=10$.

large magnetic field dependence [67]. It is shown that the main contribution to the friction coefficient can be expressed by η_2, and the friction coefficient depends on the magnetic field considerably. Since the dynamical process on the metal surface is affected by the friction force [68], we can expect that the strong magnetic field dependence comes out in dynamical process on the metal surface, for instance, the sticking coefficient.

5. Summary

In this article, first, we have described the exact solution of the single orbital Anderson model for a single impurity with the use of the Bethe Ansatz method. In addition, we discussed the thermodynamic properties of this model analytically and numerically. In second part of this article, we have mentioned the thermodynamic properties of the highly correlated degenerate Anderson model obtained by the Bethe Ansatz method. In order to investigate the heavy electron Ce compounds, the calculation has been done by including the spin-orbit coupling and also the crystalline field. The last part has been devoted to the application of the exact solution to the transport phenomena and the dynamical properties by means of the local Fermi-liquid theory developed in the impurity problem.

Acknowledgement

I am indebted to Dr. N. Kawakami for his helpful discussions and for his comments on the manuscript. Most of the work presented here is a collaboration with my colleagues, Drs. N. Kawakami, H. Kasai, A. Nakamura and T. Usuki. I should like also to thank Miss. K. Fukushima for her assistance with the typing of the manuscript.

References

1. P.W. Anderson: Phys. Rev. 124, 41 (1961)
2. See, e.g. Crystalline Field and Anomalous Mixing Effects in f-electron Systems, ed. by T. Kasuya (North-Holland, Amsterdam 1985)
3. See, e.g. Springer Series in Solid-State Sciences 59 (1985)
4. H.R. Krishna-murthy, J.W. Wilkins and K.G. Wilson: Phys. Rev. B21, 1003, 1044 (1980)
5. A.M. Tsvelick and P.B. Wiegmann: Adv. Phys. 32, 453 (1983)
6. A. Okiji and N. Kawakami: J. Appl. Phys. 55, 453 (1984);
   Springer Series in Solid-State Sciences 62, 46 (1985)
7. K.G. Wilson: Rev. Mod. Phys. 47, 773 (1975)
   N. Andrei, K. Furuya and J.H. Lowenstein: Rev. Mod. Phys. 55, 331 (1983)
8. C.N. Yang: Phys. Rev. Lett. 19, 1312 (1967)
9. P.B. Wiegmann: Phys. Lett. 80A, 163 (1980)
10. N. Kawakami and A. Okiji: Phys. Lett. 86A, 483 (1981)
11. N. Kawakami and A. Okiji: J. Phys. Soc. Jpn. 51, 1145, 2043 (1982)
    A. Okiji and N. Kawakami: J. Phys. Soc. Jpn. 51, 3192 (1982)
    P.B. Wiegmann and A.M. Tsvelick: J. Phys. C16, 2281 (1983)
12. A. Okiji and N. Kawakami: Solid State Commun. 43, 365 (1982)
    P.B. Wiegmann, V.M. Filyov and A.M. Tsvelick: Sov. Phys. JETP Lett. 35, 92 (1982)
13. K. Yosida and K. Yamada: Prog. Theor. Phys. Suppl. 46, 244 (1970)
    K. Yamada: Prog. Theor. Phys. 53, 970 (1975)
14. See, e.g. K. Yosida and A. Yoshimori: Magnetism V, ed. by H. Suhl (Academic Press, New York 1973) p.253
15. K. Ueda and W. Apel: J. Phys. C16, L849 (1983)
16. V. Zlatic and B. Horvatic: Phys. Rev. B28, 6904 (1983)
17. H. Ishii and K. Yosida: Prog. Theor. Phys. 38, 61 (1967)
18. J.R. Schrieffer and P.A. Wolff: Phys. Rev. 149, 491 (1966)
19. N. Andrei: Phys. Rev. Lett. 45, 379 (1980)
    P.B. Wiegmann: J. Phys. C14, 1463 (1981)
20. N. Kawakami and A. Okiji: Solid State Commun. 43, 467 (1982)
    V.M. Filyov, A.M. Tsvelick and P.B. Wiegmann: Phys. Lett. 89A, 157 (1982)
21. See, e.g. H.B. Thacker: Rev. Mod. Phys. 53, 253 (1981)
22. A.M. Tsvelick and P.B. Wiegmann: Phys. Lett. 89A, 368 (1982);
    J. Phys. C16, 2321 (1983)
    N. Kawakami and A. Okiji: J. Phys. Soc. Jpn. 52, 1119 (1983)
23. A. Okiji and N. Kawakami: Phys. Rev. Lett. 50, 1157 (1983)
24. N. Kawakami and A. Okiji: Phys. Lett. 98A, 54 (1983); Phys. Rev. Lett. 51, 2011 (1983)
25. A.M. Tsvelick and P.B. Wiegmann: J. Phys. C15, 1707 (1982)
26. N. Kawakami and A. Okiji: Springer Series in Solid-State Sciences 62, 57 (1985)
    P. Schlottmann: Springer Series in Solid-State Sciences 62, 68 (1985)
    A. Okiji and N. Kawakami: J. Magn. Magn. Mat. 54-57, 327 (1986)
27. A.C. Hewson and J.W. Rasul: J. Phys. C16, 6799 (1983)
28. V.T. Rajan: Phys. Rev. Lett. 51, 308 (1983)
29. P. Schlottmann: Phys. Rev. Lett. 50, 1697 (1983); Z. Phys. B51, 49 (1983)
30. N. Kawakami, S. Tokuono and A. Okiji: J. Phys. Soc. Jpn. 53, 51 (1984)
31. N. Kawakami and A. Okiji: Phys. Lett. 103A, 205 (1984)
32. P. Schlottmann: Z. Phys. B57, 23 (1984)
33. D.M. Newns and A.C. Hewson: J. Phys. F10, 2429 (1980)
34. P. Schlottmann: Z. Phys. B55, 293 (1984); Phys. Rev. B30, 1545 (1984)
35. N. Kawakami and A. Okiji: J. Phys. Soc. Jpn. 54, 685 (1985)
36. N. Kawakami and A. Okiji: J. Magn. Magn. Mat. 52, 220 (1985)
37. I. Okada and K. Yosida: Prog. Theor. Phys. 49, 1483 (1973)
38. A. Ogawa and A. Yoshimori: Prog. Theor. Phys. 53, 315 (1975)

39. K. Yamada, K. Yosida and K. Hanzawa: Prog. Theor. Phys. $\underline{71}$, 450 (1984)
40. A. Yoshimori: Prog. Theor. Phys. $\underline{55}$, 67 (1976)
    P. Nozieres and A. Blandin: J. Physique $\underline{41}$, 193 (1980)
41. B. Sutherland: Phys. Rev. Lett. $\underline{20}$, 98 (1967)
42. E. Ogievetski, A.M. Tsvelick and P.B. Wiegmann: J. Phys. $\underline{C16}$, L797 (1983)
43. H. Lustfeld and A. Bringer: Solid State Commun. $\underline{28}$, 119 (1978)
44. F. Steglich, C.D. Bredl, W. Lieke, U. Rauchswalbe and G. Sparn: Physica $\underline{126B}$, 82 (1984)
45. N. Kawakami and A. Okiji: Phys. Lett. $\underline{A115}$, 233 (1986)
46. K. Hanzawa, K. Yamada and K. Yosida: J. Magn. Magn. Mat. $\underline{47}$, 357 (1985)
47. A. Okiji and N. Kawakami: J. Magn. Magn. Mat. $\underline{54-57}$, 327 (1986)
48. K. Yosida and K. Yamada: Prog. Theor. Phys. $\underline{53}$, 1286 (1975)
49. P. Nozières: J. Low Temp. Phys. $\underline{17}$, 970 (1975)
50. D.M. Newns, A.C. Hewson, J.W. Rasul and N. Read: J. Appl. Phys. $\underline{53}$, 7877 (1982)
51. N. Kawakami and A. Okiji: Phys. Lett. $\underline{118A}$, 301 (1986)
52. D.C. Langreth: Phys. Rev. $\underline{150}$, 516 (1966)
53. F.D.M. Haldane: Phys. Rev. Lett. $\underline{40}$, 416 (1978)
54. N. Kawakami, T. Usuki and A. Okiji: J. Phys. Soc. Jpn. $\underline{56}$, 1539 (1987)
55. J. Kondo: Prog. Theor. Phys. $\underline{34}$, 372 (1965)
56. N. Kawakami and A. Okiji: Jpn. J. Appl. Phys. $\underline{26}$, 499 (1987)
57. P. Schlottmann: J. Magn. Magn. Mat. $\underline{47-48}$, 367 (1985)
58. N.B. Brant and V.V. Moshchalkov: Adv. Phys. $\underline{33}$, 373 (1984)
59. A. Nakamura, N. Kawakami and A. Okiji: J. Phys. Soc. Jpn. $\underline{56}$, 3667 (1987); Jpn. J. Appl. Phys. $\underline{26}$, 501 (1987)
60. H. Shiba: Prog. Theor. Phys. $\underline{54}$, 967 (1975)
61. A.M. Muraui, K. Knorr, K.H.J. Buschow, A. Benoit and J. Floquet: Solid State Commun. $\underline{36}$, 523 (1980)
    S. Horn, E. Holland-Moritz, M. Loewenhanpt, F. Steglich, H. Schener, A. Benoit and J. Floquet: Phys. Rev. $\underline{B23}$, 3171 (1981)
62. Y. Kitaoka, H. Arimoto, Y. Kohori and K. Asayama: J. Phys. Soc. Jpn. $\underline{54}$, 3236 (1985)
    Y. Kitaoka, K. Fujiwara, Y. Kohori, K. Asayama, Y. Ōnuki and T. Komatsubara: J. Phys. Soc. Jpn. $\underline{54}$, 3686 (1985)
    M.J. Lysak and D.E. MacLaughlin: Phys. Rev. $\underline{B31}$, 6963 (1985)
63. E.G. d'Agliano, P. Kumar, W. Schaich and H. Suhl: Phys. Rev. $\underline{B11}$, 2122 (1975)
64. A. Nourtier: J. Phys. (Paris) $\underline{38}$, 479 (1977)
65. K.P. Bohnen, M. Kiwi and H. Suhl: Phys. Rev. Lett. $\underline{34}$, 1512 (1975)
66. A. Yoshimori and J.L. Motchane: J. Phys. Soc. Jpn. $\underline{51}$, 1826 (1982)
67. A. Okiji and H. Kasai: Suface Sci. $\underline{188}$, L717 (1987)
68. P.B. Visscher: Phys. Rev. $\underline{B14}$, 347 (1977)

Fermi Surface Effects in Atom and Molecule Surface Scattering

D.M. Newns[1], *K. Makoshi*[2], *and R. Brako*[3]

[1]IBM T.J. Watson Research Center, Box 218, Yorktown Heights,
NY 10598, USA
[2] Faculty of Engineering Science, Osaka University,
Osaka 560, Japan
[3]Institute Ruder Boscovic, 41001 Zagreb, Yugoslavia

1. Introduction

This paper deals with various kinds of Fermi surface effects in the scattering of atoms and molecules from metal surfaces. At low projectile velocities, the one-body potential of the projectile leads to a coupling to electron-hole pairs which is infra-red singular, as first pointed out by Muller-Hartman et al. [1] . In practice, this phenomenon is hard to observe because the center of mass degree of freedom of atoms or molecules scattering from surfaces is so strongly coupled to the phonon system; inelastic effects are dominated by phonons and electron-hole pair excitation is secondary.

However, the internal vibrational degree of freedom of strongly bound molecules (e.g. H_2, CO, NO) is very poorly coupled to the phonon system. This is because its frequency ω_0 is of order 10 x the Debye frequency, requiring $\sim$ 10 phonons to excite it, and because at low energies the collision time τ of the molecule with the surface is long, so that the Massey-type factor $\exp(-\omega_0\tau)$ quenches vibrational excitation. Then provided the electronic coupling between the molecule and surface is adequate, energy transfer between the molecule's vibrational degree of freedom and the electron-hole degrees of freedom in the surface will be observable. This energy transfer seems indeed to have been observed for NO on Ag(111) [2,3].

At high projectile energies and grazing incidence conditions, another type of Fermi surface phenomenon comes into play. The component of projectile velocity parallel to the surface $V_\parallel$, can be thought of as inducing a Doppler shift in the metal Fermi surface, which in the jellium model is displaced in k-space by $mV_\parallel$, where m is electron mass. This displacement leads to a smearing of the Fermi distribution by an effective temperature $T \sim k_F V_\parallel$, which can result in nonzero occupation of projectile valence orbitals lying above the Fermi level, and normally empty. Furthermore, in the case of degenerate shells, such as the H n = 2 or n = 3 shells, the density matrix of the valence orbitals becomes characteristically asymmetric, leading to measurable polarization of light emitted when the excited species radiatively relaxes, a fingerprint which seems to be broadly understood by theory [4].

2. Atom-Surface Scattering at Low Energy

2.1 Formulation

It was first pointed out by Muller-Hartman et al. for scattering of a projectile from a metal surface [1] that if the trajectory approximation is used, when the center of the mass of the projectile is treated classically, then the surface sees a slowly switched on and off time-dependent local perturbation: Such a perturbation surprisingly returns the metal to its ground state with a probability that does not tend to unity in the slow limit. This is an example where the adiabatic theorem fails, duc to the infra-red singular spectrum of electron-hole excitations coupled to a local perturbation. Associated with this electronic Debye-Waller factor problem, is an energy loss to the electron-hole pair excitations of the substrate. The energy loss itself vanishes in the slow limit. Let us consider how these phenomena may be calculated.

In calculating the energy transfer between the translational degree of freedom of the projectile and the substrate electron-hole pairs, the trajectory approximation has been used to treat the strong coupling case [5-7], while essentially the Born approximation is used to treat the weak-coupling case quantum-mechanically [8,9]. Since strong coupling is needed to get significant electron-hole inelasticity, and the semiclassical approaches are more transparent, we shall here discuss the trajectory approximation (TA) approach.

Now in the trajectory approximation the centre of mass of the projectile is treated classically and the electronic matrix elements are a function of the position of the projectile and thus of time t along the trajectory. The Hamiltonian in the TA is a time-dependent one in the subspace of the electronic states only, which may be written

$$H = H_0 + V(t) , \quad \text{where} \tag{1}$$

$$H_0 = \sum_k \epsilon_k C_k^+ C_k \tag{2}$$

is the Hamiltonian of the clean metal surface. Here $| k >$ is an eigenstate of the clean metal surface, with energy eigenvalue ϵ_k, treated in single-particle approximation. For the moment, we neglect spin. The perturbation $V(t)$ represents the effect of an atom at time t on its trajectory over the surface.

Now to get a significant coupling to electron-hole pairs, it turns out that a resonance in the projectile orbitals is necessary. For consistency in this paper, we shall model this resonance using the non-interacting Anderson model

$$V(t) = \epsilon_a(t)n_a + \sum_k [V_k(t)C_k^+ C_a + \text{h.c.}] . \tag{3}$$

In (3), C_a represents the valence orbital $| a >$ on the projectile of energy ϵ_a, for the moment assumed nondegenerate, which is coupled to the metal state $| k >$ via a hopping matrix element V. Both ϵ_a and V vary with distance from the surface and thus with time t. When the atom is far from the surface, (3) describes a valence level at ϵ_a. As the atom approaches the surface, the level acquires a lifetime broadening $\Delta \sim \pi V_k^2 \rho$, where ρ is the DOS of the metal substrate; i.e. we have a resonance at ϵ_a of width Δ.

The quantity we wish to calculate is the probability $P(\epsilon)$ of transfering energy ϵ during the collision,

$$P(\epsilon) = < \infty | \delta(\epsilon - [H(\infty) - E_0]) | \infty > , \tag{4}$$

where $H(\infty)$ is the Hamiltonian far from the surface and E_0 the ground state energy. Introducing the Fourier transform $p(t)$ of $P(\epsilon)$

$$p(t) = < \infty | \exp [- i[H(\infty) - E_0]t] | \infty > , \tag{5}$$

$p(t)$ may be written in Heisenberg representation

$$p(t) = < t_0 | \exp [-iH_H(\infty)t] | t_0 > \exp [iE_0 t], \quad \text{where} \tag{6}$$

$$H_H(\infty) = U^+(\infty, t_0)H(\infty)U(\infty, t_0) . \tag{7}$$

In (6-7) t_0 is the initial time before scattering, and $U(\infty, t_0)$ the evolution operator.

In Eq. (6-7), the problem separates into two halves. First, we have to calculate the time-independent Hamiltonian $H_H(\infty)$, a problem defined in the subspace of one particle. Then, the expectation value (6) is to be calculated. In this paper, we adopt the relatively conventional approach of Ref. 6, rather than the unconventional time-dependent Tomonagon approach of Ref. 5, or the tour de force of Makoshi's Keldysh treatment [7].

<u>2.2 One-body Hamiltonian in slow limit</u>

First we write down the equations of motion for the operators C_a, C_k

$$i\dot{C}_a = \varepsilon_a(t)C_a + \sum_k V_k^*(t)C_k \tag{8}$$

$$i\dot{C}_k = \varepsilon_k C_k + V_k(t)C_a . \tag{9}$$

Eq. (9) can be solved for C_k and inserted into (8). Then (8) can be solved if we make two additional assumptions; that $V_k(t)$ has the separable form $u(t)V_k$, and that the a-level lifetime broadening $\Delta(t)$, defined by

$$\Delta(t) = \pi u^2(t) \sum_k V_k^2 \delta(\varepsilon - \varepsilon_k) \tag{10}$$

varies little with energy ε. This is a reasonable assumption if the metal conduction band is wide on the scale of Δ. We find that

$$\begin{aligned}
C_a(t) = {}& -i\int_{t_0}^{t} dt' \exp\left[\int_{t}^{t'} Z(\tau)d\tau\right]u(t') \\
& \times \sum_k V_k^* \exp[-i\varepsilon_k(t'-t_0)]C_k^0 \\
& + C_a^0 \exp\left[-\int_{t_0}^{t} Z(\tau)d\tau\right], \qquad \text{where}
\end{aligned} \tag{11}$$

$$Z(\tau) = \Delta(\tau) + i\varepsilon_a(\tau) \tag{12}$$

and $C^0 = C(t_0)$. Hence substituting (11) into (9), and solving the first order differential equation for C_k, we obtain

$$\begin{aligned}
\exp(i\varepsilon_k t)C_k(t) = {}& \sum_{k'} C_{k'}^0 \Big[\delta_{kk'} - \int_{t_0}^{t} dt' V_k(t') \exp(i\varepsilon_k t') \\
& \times \int_{t_0}^{t'} dt'' V_{k'}(t'') \exp(-i\varepsilon_{k'}t'') \exp\left(-\int_{t''}^{t'} Z(\tau)d\tau\right)\Big] \\
& - iC_a^0 \int_{t_0}^{t} dt' V_k(t') \exp(i\varepsilon_k t') \exp\left(-\int_{t_0}^{t'} Z(\tau)d\tau\right).
\end{aligned} \tag{13}$$

Now using the fact that $\exp[-\int_{t''}^{t'} Z d\tau]$ is large only for $(t'-t'') < \Delta^{-1}(t')$, this factor can be replaced by $\exp[-(t''-t')Z(t')]$. Then the second term in (13) vanishes, and the first one gives

$$e^{i\epsilon_k t}C_k(t) = \frac{1}{2\pi} \int d\epsilon' C^0_{\epsilon'} \int_{t_0}^{t} dt' \exp\left[i(\epsilon_k - \epsilon')t' - 2i\delta(\epsilon', t')\right], \tag{14}$$

where the phase shift δ is

$$\delta(\epsilon, t) = \tan^{-1} \frac{\Delta(t)}{\epsilon - \epsilon_a(t)} \;.$$

δ is the phase shift at energy ϵ seen by the metal electrons due to their interaction with the projectile orbital at time t along the trajectory.

$H_H(\infty)$ may now be obtained by substituting (14) into (1) and taking the limit $t \rightarrow \infty$ (where V (t) vanishes), giving

$$H_H(\infty) = H_0 + H' , \quad \text{where} \tag{15}$$

$$H' = \sum_{kk'} W_{kk'} C_k^+ C_{k'} \quad \text{and} \tag{16}$$

$$W_{kk'} = \frac{1}{\pi \rho_k} \int dt \dot{\delta}(\epsilon_k, t) \exp\left[i(\epsilon_k - \epsilon_{k'})t\right] \;. \tag{17}$$

Here $\rho_k = \rho(\epsilon_k) \simeq \rho(\epsilon_{k'})$ is the DOS at ϵ_k (ϵ_k assumed near $\epsilon_{k'}$). Eq.'s (15-17) describe a time-independent Hamiltonian in which the scattering potential W is controlled by the rate of change of phase shift δ in the a-orbital. If δ changes slowly (slow motion of the projectile) then only states of nearby energy $\epsilon_k \simeq \epsilon_{k'}$ enter into W. This property is independent of the strength of the potential, which controls the magnitude of δ, which however is limited to π by unitarity.

2.3 Fermi statistics

Now when we substitute (15) into (6), Fermi statistics will allow only transitions from filled to empty states, and since $\epsilon_k \simeq \epsilon_{k'}$, this will restrict us to states close to ϵ_F. The model now has all its k-dependence within ϵ_k, and is an effective one-dimensional one (actually its dimension is along the trajectory time axis due to the elimination of space dimensions by the restriction to a single phase shift δ) :

$$H_H(\infty) = \sum_k v_F k C_k^+ C_k + \frac{2v_F}{L} \sum_{k, q} w(q) C_k^+ C_{k-q} \quad \text{where} \tag{18}$$

$$w(q) = \int dt \dot{\delta}(t) \exp\left[iv_F qt\right] . \tag{19}$$

In (18), we formally introduce a Fermi velocity by $\epsilon_k = v_F k$ and define $\rho = L/2\pi v_F = \rho(\epsilon_F)$.

The standard procedure for one-dimensional models of this type, which are exactly soluble, is Luttinger (or Tomonaga) bosonisation [10]. We introduce bosons

$$b_p = \sqrt{\frac{2\pi}{pL}} \sum_k C_k^+ C_{k-p} \quad ; p > 0$$

$$b_p^+ = \sqrt{\frac{2\pi}{pL}} \sum_k C_{k-p}^+ C_k \quad ; p > 0, \tag{20}$$

in terms of which $H_H(\infty)$ becomes

$$H_H(\infty) = v_F \sum_{p>0} p b_p^+ b_p + \left(\frac{2}{L\pi}\right)^{1/2} v_F \sum_{p>0} p^{1/2} (w_p^* b_p + w_p b_p^+) . \qquad (21)$$

Equation (21) is a driven harmonic oscillator form in terms of which (6) can be calculated exactly [10]. We get [5-7] at zero temperature

$$p(t) = \exp\left[\pi^{-2} \int_0^\infty d\omega\, \omega \mid \delta(\omega) \mid^2 (e^{-i\omega t} - 1) \right], \qquad \text{where} \qquad (22)$$

$$\delta(\omega) = \int dt\, \delta(t) e^{i\omega t}\, dt . \qquad (23)$$

Note that in calculating (16) to order $\dot\delta$, a constant "relaxation shift" term of order $(\dot\delta)^2$ was omitted; it has been retrieved to cancel out an unphysical relaxation shift otherwise present in (22).

2.4 Application of result

Before we consider any application of the result (22), it needs some generalization. So far, the spin label σ has been neglected. There may also be other degeneracies present, such as angular momentum m about the surface normal. Secondly, we generalize to finite temperature [5-7], giving

$$p(t) = \exp\left[\pi^{-2} \sum_{m,\sigma} \int_0^\infty d\omega\, \omega \mid \delta_{m\sigma}(\omega) \mid^2 \left\{ \coth \frac{\beta\omega}{2} (\cos \omega t - 1) - i \sin \omega t \right\} \right], \quad (24)$$

where $\beta = 1/kT$, and $\delta_{m\sigma}$ is phase shift of the m, σ channel.

Let us first consider the Debye-Waller factor P_0, which is the coefficient of the delta function $\delta(\epsilon)$ in $P(\epsilon)$. It is obtained from the $t \to \infty$ limit of p (t) as

$$P_0 = \exp\left[-\pi^{-2} \sum_{m,\sigma} \int_0^\infty d\omega\, \omega \mid \delta_{m\sigma}(\omega) \mid^2 \coth \frac{\beta\omega}{2} \right]; \qquad (25)$$

P_0 is seen to decrease as T increases, in the usual manner of the Debye-Waller factor.

When $T = 0$, (25) becomes

$$P_0 = \exp\left[-\pi^{-2} \sum_{m,\sigma} \int_0^\infty d\omega\, \omega \mid \delta_{m\sigma}(\omega) \mid^2 \right]. \qquad (26)$$

Eq. (26) has the property that if ω is rescaled by a factor $\omega \to \alpha\omega$, it is invariant. In other words, we retrieve the result of Muller-Hartman' et al. [1], that the ground state $\to$ ground state amplitude is less than unity whatever the time scale τ of the projectile motion. The result will apply at low temperatures $kT < < \tau^{-1}$.

A model for time variation of the phase shift $\delta(t)$ might be (in each channel)

$$\delta(t) = \frac{\delta^0}{1 + t^2/\tau^2} , \qquad (27)$$

when we obtain $P_0 = \exp (- w)$, with

$$w = \frac{1}{4} \sum_{m,\sigma} (\delta^0_{m,\sigma})^2 . \tag{28}$$

It is also convenient to think in terms of the maximum change in the number of electrons in a given channel, $n_{max,\,m\sigma} = \delta^0_{m\sigma}/\pi$. In Table 1 , we make some very naive estimates of Debye-Waller factors for some common atoms and molecules. It is seen that substantial effects (i.e. small P_0) may be obtained in suitable cases, e.g. O_2 going to $O_2^{\parallel}$ at the surface. The electron hole scattering should act to reduce quantum-mechanical coherence at the surface, and this indeed seems to be occurring in NO scattering off Ag (111) [11].

The mean energy loss by the projectile in the TA is always positive and temperature-independent

$$\overline{E} = \sum_{m,\sigma} \pi^{-1} \int_{-\infty}^{\infty} dt \, (\dot{\delta}_{m\sigma}(t))^2 . \tag{29}$$

In the model (27), we obtain

$$\overline{E} = \frac{\pi^2}{4} \tau^{-1} \sum_{m\sigma} n^2_{max,\,m\sigma} . \tag{30}$$

$\overline{E}$ is typically of order τ^{-1}.

A numerical evaluation of $P(\epsilon)$ using the model (27) is easily done, and the results are given for various temperatures in Fig. 1.

Figure 1 shows the typical evolution from low temperature quantum behaviour, dominated by energy loss with a significant DW factor, to high-temperature classical behaviour which is Gaussian-like around $\epsilon = \overline{E}$, with negligible DW factor.

An ingenious analytically soluble case was found by Gunnarsson and Schönhammer [5] for the spinless case

Table 1. Estimates of coherently scattered fraction at T=0 for various atoms and molecules

Particle		At T=0 w	P_0
H, Li	$n_{max\,\uparrow} = -0.5 = -n_{max\,\downarrow}$	1.25	0.3
NO Axis normal to surface	$n_{max\,\uparrow\,1} = -0.5 = -n_{max\,\downarrow\,1}$ $n_{max\,\uparrow\,-1} = n_{max\,\downarrow\,-1} = 0$	1.25	0.3
O_2 Axis normal to surface	$n_{max\,\uparrow\,1} = n_{max\,\uparrow\,-1} = -0.5$ $= -n_{max\,\downarrow\,1} = -n_{max\,\downarrow\,-1}$	2.5	0.1
O_2 Axis parallel to surface	$n_{max\,\uparrow\,1} = n_{max\,\uparrow\,-1} = -0.5$ $= n_{max\,\downarrow\,1} = n_{max\,\downarrow\,-1}$	5.0˙	.01
H_2 Axis parallel to surface	$n_{max\,\sigma g} \simeq -n_{max\,\sigma u} = \left\{ \begin{array}{l} 0.1 \\ 0.3 \end{array} \right.$	0.1 0.9	0.9 0.6

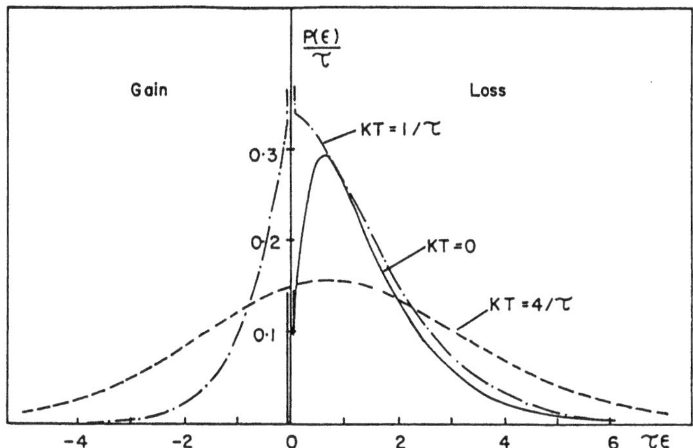

Fig. 1. The loss function $P(\varepsilon)$ at three different temperatures T. The phase shift of the perturbation has a Lorentzian time dependence with a characteristic time τ, equation (27). The intensity of the no-loss lines are given by $P_0 = 0.41$ at $kT = 0$, $P_0 = 0.02$ at $kT = 1/\tau$ and $P_0 \simeq 0$ at $kT = 4/\tau$.

$$\dot{\delta}(t) = \frac{\Delta t/2}{t^2 + (\Delta t/2)^2} - \frac{\Delta t/2}{(t + T)^2 + (\Delta t/2)^2} . \tag{31}$$

In this model, the initially empty level crosses through the Fermi level between $-\Delta t/2$ and $+\Delta t/2$, then spends time T below ε_F, finally crossing back through ε_F between $T - \Delta t/2$ and $T + \Delta t/2$. For (31) all the integrals in (22) may be performed [5], yielding

$$P(\varepsilon) = \frac{(\Delta t)^2}{(\Delta t)^2 + T^2} \left(\delta(\varepsilon) + T^2 \varepsilon e^{-\varepsilon \Delta t} \right) . \tag{32}$$

The result (32) is closely related to an earlier one derived by Norskov using a semiclassical argument in the limit $\Delta t << T$. He considers the level $\varepsilon_a(t)$ dropping below the Fermi level ε_F at $t = 0$, but being filled at a rate $2\Delta(t)$, the survival probability of the hole to depth ε being

$$\exp\left[-2 \int_0^{\varepsilon_1/\dot{\varepsilon}_a} \Delta(t) dt \right] \simeq \exp\left[-2\Delta\varepsilon_1/\dot{\varepsilon}_a \right] , \tag{33}$$

treating $\dot{\varepsilon}_a$ and Δ as constant near ε_F. On emerging from the Fermi level, the electron will similarly survive in the level to ε_2 with probability $\exp[-2\Delta\varepsilon_2/\dot{\varepsilon}_a]$. Hence the probability $P(\varepsilon)$ of an electron-hole pair of energy $(\varepsilon_2 - \varepsilon_1) = \varepsilon$ being excited is seen to be

$$P(\varepsilon) = \varepsilon \left(\frac{2\Delta}{\dot{\varepsilon}_a} \right)^2 \exp\left[\frac{-2\Delta\varepsilon}{|\dot{\varepsilon}_a|} \right] , \tag{34}$$

which agrees with (32) in the $T >> \Delta t$ limit if we identify $\Delta t \equiv 2\Delta/|\dot{\varepsilon}_a|$.

2.5 Effectiveness of electron-hole mechanism

The electron-hole mechanism is not an especially effective mechanism in coupling to the center of mass degree of freedom of projectiles. In the case of rare gas atoms, such as He at thermal

105

energies, numerical estimates of the energy transfer have been made [9,12], and it is found to be exceedingly small. This is not surprising since only small phase shifts are generated when rare gas atoms interact with metal surfaces.

To get a significant electronic effect, a strong electronic interaction with the surface concentrated in a few channels is needed - the effect produced by long-range dipole-image forces for example is relatively weak [13]. In other words, chemical interaction such as occurs between H, H_2, O_2, CO and NO and suitable surfaces should lead to substantial phase shifts. The problem is that in such cases the projectile develops a potential well .5-2 eV deep, which accelerates a thermal particle to energies of 5000 - 20, 000 K. The impact with the surface is strong enough to excite in many cases more phonons than electron-hole pairs, masking the electron-hole effect.

If, however, we go to systems such as NO or O_2, which on account of their nonzero spin should have a relatively strong electron-hole effect (especially O_2 - see Table 1), we should expect significant effects on the Debye-Waller factor. When careful surface scattering calculations are done including phonons, it will then be found that experimental data show less sign of coherent effects than the theory predicts. This is precisely the situation encountered by Brenig and coworkers for NO on Ag(111) [11], and it should be still stronger for O_2. It seems likely that surface scattering experiments combined with careful calculations including phonons will reveal electron-hole excitation to an increasing extent in the future.

3. Vibrational Excitation in Molecule-Surface Scattering

3.1 Introductory

In this section we shall consider the scattering of molecules having a high intramolecular vibration frequency from metal surfaces at thermal and low hyperthermal energies. A remarkable fact is that in this regime the molecule's internal degree of freedom is very weakly coupled to the solid by mechanical or phononic effects [14]. The time τ of scattering from any repulsive wall model of the atom-surface potential is much longer than the vibrational time ω_0^{-1} of a molecule of vibrational frequency ω_0. We are discussing systems like NO or CO whose vibration frequencies lie in the 200-300 meV range.   Hence the Massey factor $\exp(- \omega_0\tau)$, which comes into the probability of vibrational excitation by the direct collision process, is negligible. Furthermore, a number of target phonons of order 10 is required to excite the molecular vibration degree of freedom. Such a 10-phonon process is highly improbable, and may be ignored in the following.

Hence the transfer of energy between the molecule's vibrational degree of freedom and a metal surface proceeds through the electronic degrees of freedom.   In gas-phase electron-molecule scattering it is well known that strong excitation of the vibrational degree of freedom of the molecule occurs for beam energies of a few eV [15]. At surfaces, the lifetime broadening of, for example, the 250 meV vibration of chemisorbed CO is considered to be of electronic origin [16], though it is very small.   The mechanism believed to operate [16] is that vibration of the CO bond shifts the π^* orbital in energy; a contraction of the molecule raises the π^* and a dilation lowers it.   The π^* orbital has an intrinsic lifetime broadening Δ, resulting in a Lorentzian density of states in the orbital overlapping the Fermi energy at some point.   Dissipation of the CO vibrational energy then occurs via excitation of electron-hole pairs in the Lorentzian DOS when it is driven to oscillate in phase with the vibrational degree of freedom.

The electron-hole degree of freedom in molecule-surface scattering was considered by Gadzuk et. al. [17], who essentially discussed a non-adiabatic effect when a narrow molecular affinity antibonding level crosses the Fermi level in a downward sense, and then again in an upward sense, during scattering from the surface. If the level width Δ is narrow, relative to the time to cross the Fermi level, then the outgoing and ingoing path of the system may be asymmetric, resulting in vibrational excitation of the molecule.

In this paper, we shall discuss a specific theoretical approach based on an experimental observation of vibrational excitation in the scattering of NO from Ag (111) [2]. The result has one distinctive feature not present in the theoretical approach of Ref. 17, in that the vibrational excitation probability p_1 is strongly surface-temperature dependent. In fact it approximately obeys the Arrhenius law

$$p_1 = \exp(-\omega_0/ kT_s) \, ,$$

where T_s is surface temperature. The approach that we shall describe is principally an adiabatic one.

3.2 Model Hamiltonian

The electron-vibration coupling we use is based on the gas-phase PE curves of NO and NO^-, which show that NO^- has an extended bond distance due to the additional occupation of the π^* orbital. The approximate model for the NO Hamiltonian is then

$$h_{NO} = \omega_0 b^+ b + \lambda(b^+ + b) n_a \, . \tag{35}$$

In (35) we have bosonized the oscillator, of frequency ω_0 , and introduced its displacement, proportional to $(b^+ + b)$. The notation $| a >$ is used for the π^* orbital, to conform with that of section 2, and n_a is its occupation. By comparing the oscillator extension in (35) at $n_a = 1$ with the experimental bond length difference between NO and NO^- [18], we find $\lambda = 0.32$ eV. A simplification made in (35) is the neglect of softening in the vibrational frequency ω_0 on negative ionization of NO.

The electronic part of the Hamiltonian for NO interacting with a surface is taken to be given by (1-3), as discussed in the foregoing sections. Finally we assume the Trajectory approximation in the center of mass coordinate of the NO molecule. The resulting Hamiltonian is

$$H(t) = \epsilon_a(t)n_a + \sum_k \epsilon_k n_k + \sum_k [V_k(t)C_k^+ C_a + \text{h.c.}] + \omega_0 b^+ b + \lambda(b^+ + b) n_a \, . \tag{36}$$

Our approach is to treat the rotational degrees of freedom of the NO molecule in an average way. They do not enter into (36) explicitly. but $V_k(t)$ is to be interpreted as an orientational average matrix element.

3.3 Excitation Probability

The number of bosons $\bar{n}$ in the oscillator at $t = + \infty$, which we take to be p_1 since the two-quantum excitation probability of the oscillator is negligible, may be defined as

$$\bar{n} = < \infty | b^+ b | \infty > \, , \tag{37}$$

where $| \infty >$ is the mixed quantum state at $t = \infty$. By making the unitary transformation (7), (37) may be rewritten

$$\bar{n} = < t_0 | b^+(\infty)b(\infty) | t_0 > \, , \tag{38}$$

where now $| t_0 >$ is as before the initial wavefunction, and $b(t)$ is the boson operator in Heisenberg representation at time t. $b(t)$ obeys the equation of motion

$$ib = \omega_0 + \lambda n_a \quad , \tag{39}$$

with solution (defining $b^0 = b(t_0)$)

$$b(t) = -i\lambda e^{-i\omega_0 t} \int_{t_0}^{t} e^{i\omega_0 t'} n_a(t')\, dt' + e^{i\omega_0(t-t_0)} b^0 \quad . \tag{40}$$

Substituting (40) into (38), and assuming $< t_0 \mid b^{0+} b^0 \mid t_0 >$ is zero (oscillator initially in ground state), then (38) becomes the Fourier transform of a correlation function:

$$\bar{n} = \lambda^2 \int_{t_0}^{\infty} dt_1 \int_{t_0}^{\infty} dt_2 e^{-i\omega_0(t_1 - t_2)} g(t_1, t_2) \quad , \tag{41}$$

where the correlation function g is

$$g(t_1, t_2) = < t_0 \mid n_a(t_1) n_a(t_2) \mid t_0 > \quad . \tag{42}$$

In general the evaluation of (42) is very difficult, but we shall here assume that the smallness of $\bar{n}$ (maximum .07 in the experiment) justifies a Born approximation treatment. Working then to order λ^2 in (41), $n_a(t)$ in (42) may be calculated as if $\lambda = 0$. We may then return to the equation of motion for $C_a(t)$ [8-9], and its solution (11) in the wide band limit.

Substituting (11) into (42), we obtain three relatively complicated terms

$$g = g_1 + g_2 + g_3 \quad . \tag{43}$$

Analysis showed [3], that the only one of these three terms which seemed to be appreciable in the region of experimental interest

$$\tau^{-1} < kT < \omega_0 < \Delta, \epsilon_a \tag{44}$$

was the term g_2, found to be

$$g_2(t_1, t_2) = (\Delta_0/\pi)^2 \int d\epsilon f(\epsilon) \int dE(1 - f(E))$$
$$F^*(t_1, \epsilon) F(t_2, \epsilon) F(t_1, E) F^*(t_2, E) \quad , \tag{45}$$

where $f(\epsilon)$ is the Fermi function

$$F(t,x) = e^{-\int_{t_0}^{t} dt\Delta(t)} \int_{t_0}^{t} d\tau u(\tau) e^{-ix\tau + \int_{t_0}^{\tau} d\tau Z(\tau)} \quad , \tag{46}$$

and Z is defined in (12).

In terms of (41), $\bar{n}$ is defined as an 8-dimensional integral. In order to evaluate it, we assume that the inequalities in (44) are essentially 'much less than' signs. Then omitting the details, which may be found in Ref. [3], $\bar{n}$ is found to be

$$\bar{n} = 2\lambda^2 \omega_0 \pi e^{-\beta\omega_0} \int_{-\infty}^{\infty} dt \rho_a^2(t) \quad . \tag{47}$$

In (47), $\rho_a(t)$ is the DOS in the Lorentzian resonance of the π^* orbital at the Fermi level; taking Fermi as energy zero, ρ_a is

$$\rho_a(t) = \pi^{-1}\Delta(t)/[\epsilon_a^2(t) + \Delta^2(t)] \quad , \qquad \text{while } \beta = 1/kT_s. \tag{48}$$

3.4 Alternative Derivation

An alternative derivation, of which we are not quite sure of the conditions of validity, is based on a master equation for the occupation $\bar{n}(t)$ of the oscillator boson at time t along the trajectory. This may be written

$$d\bar{n}/dt + \bar{n}\Gamma(t) = e^{-\beta\omega_0}\Gamma(t) \quad . \tag{49}$$

In (49), the stationary value of $\bar{n}$ is just the Boltzmann factor (ignoring two-quantum effects). At zero temperature and small times, $\bar{n}$ decays with rate Γ.

Now Γ, the rate of damping of the vibrations of a fixed oscillator on a cold surface, has been calculated already by Persson and Persson [16]. Γ is given by

$$\Gamma = 2\pi\omega_0\lambda^2\rho_a^2 \quad . \tag{50}$$

If we solve (49) for $\bar{n}$ to $O(\lambda^2)$ and substitute in (50), we retrieve our earlier expression (47-48).

3.5 Comparison with Experiment

We first of all note that (47) indeed contains the Boltzmann factor to which we drew attention in Section 3.1as a characteristic feature of the experimental data.

In order to compare (47) with the velocity dependence seen in the experimental results of Ref. 2, it is necessary to make some assumptions about the spatial variation of $\Delta(z)$ and $\varepsilon_a(z)$, and about the trajectory $z(t)$.

We assume the forms

$$\Delta = \Delta_0 e^{-2\alpha z} ,$$
$$\varepsilon_a(z) = \varepsilon_\infty - e^2/4(z - z_0) \quad . \tag{51}$$

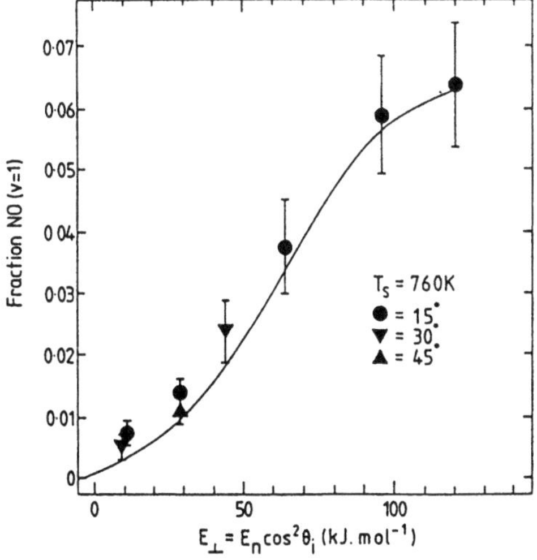

Fig. 2.   Plot of vibrational excitation probability vs. normal kinetic energy of the NO beam. Points, experiment of Ref. 2, curve, results of present theory.

The exponential form for Δ in (51) is conventional, whereas the classical image formula could not be used at any smaller distances than we employ (i.e. it could not be used at any higher energies of NO). ϵ_∞ is the vertical electron affinity of NO [18].

The potential energy curve for the NO center of mass is taken to be of Morse type

$$U(z) = D(e^{-2\alpha z} - 2e^{-\alpha z}) \ , \tag{52}$$

with α the same as in (51). The depth D is taken to be 300 meV, based on experimental data [2].

It is now possible to calculate the energy dependence of $\bar{n}$ as a function of the two parameters Δ_0 and z_0 in (51). A fairly unique fit is obtained for $\Delta_0 = .132$ eV and $z_0 = -1.93$ Å.

The comparison with experiment testing these two parameters is illustrated in Fig.2, and is seen to be good. In Fig. 3 we illustrate the behaviour of the potential energy U(z), the level variation $\epsilon_a(z)$ and the width variation $\Delta(z)$ as a function of z. The system is seen to have extremely weak electronic coupling to the substrate at the equilibrium position, i.e. to be essentially physisorbed, but nevertheless at impact the electronic coupling becomes quite strong, enabling the electron-hole excitation mechanism to operate.

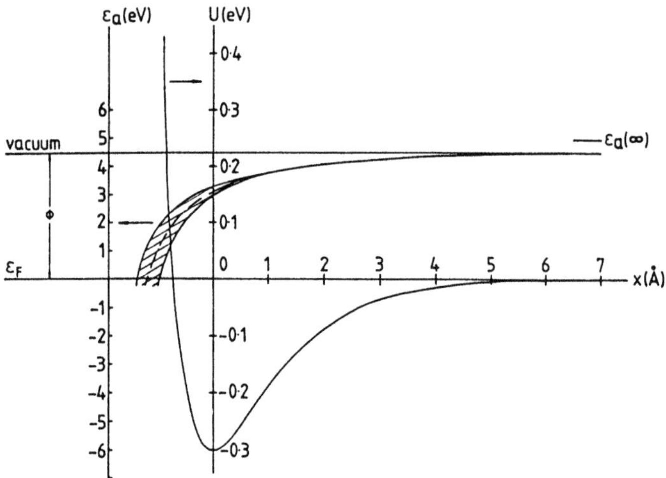

Fig. 3.   Variation of $\epsilon_a(z)$, $\Delta(z)$ and U(z) in the model used in the theoretical curve of Fig. 2. Dot-dash curve is $\epsilon_a(z)$. Hatched area is region between $\epsilon_a(z) - \Delta(z)$ and $\epsilon_a(z) + \Delta(z)$. Full curve is potential energy U(z).

4. Charge Transfer in Atom-Surface Collisions

In this section we shall turn to the problem of charge transfer between an atom and a surface resulting from collision with the surface. Compared with excitation of electron-hole pairs in the metallic substrate, or excitation of vibrational degrees of freedom in the projectile, charge transfer involves excitation of an electronic degree of freedom, which normally requires a larger amount of energy. Hence we must go to a regime of faster, higher energy projectiles to realise these charge transfer processes. Thus the field of charge transfer involves projectile energies of a few eV to a few keV.

The model we employed earlier (spinless Anderson model) provides a good elementary basis for discussing the charge transfer process. We imagine the projectile as having its valence orbital $| a >$ empty. Following collision with the metallic target, we calculate the probability $< n_a >$ that the valence orbital be occupied. This turns out to be a surprisingly rich problem, involving as it does the magnitude and direction of the velocity, the energy level variation and the lifetime broadening of the projectile and the temperature of the target.

The final charge $< n_a >$ may be written in Schrödinger picture as $< \infty | C_a^+ C_a | \infty >$, where $| \infty >$ is the wavefunction of the system at $t = \infty$. But it is more convenient to write

$$< n_a > = < t_0 | C_a^+(\infty) C_a(\infty) | t_0 > \ , \tag{53}$$

where $| t_0 >$ is the initial state and $C_a(t)$ is the operator corresponding to state $| a >$ in Heisenberg picture.

We may now substitute the expression (11) for $C_a(t)$ into (53) to give the final occupation. Before doing this, we need to take into account a velocity effect which is not associated with the explicit time dependence of the quantities ε_a and Δ in (11). Assuming a jellium (uniform positive background) model of the surface, motion of the projectile parallel to the surface will not give rise to any time dependence of ε_a or Δ, the time dependence in the phase of the V_k matrix elements having been lost. If we restore this time dependence, the result is equivalent to a simple Doppler shift in k-space of the jellium Fermi surface. This result was demonstrated by van Wunnik, Brako, Makoshi and Newns [19], and was earlier known to van Wunnik [20], and independently Bergdorfer et al. [21]. Substituting (11) into (53), and taking the $t \to \infty$ limit we have

$$< n_a > = n_{a1} + n_{a2} \ , \qquad \text{where} \tag{54}$$

$$n_{a1} = n_0 \exp \left[- 2 \int_{-\infty}^{\infty} \Delta(\tau) d\tau \right] \ , \tag{55}$$

and (taking Fermi level as energy zero)

$$n_{a2} = (4\pi^2)^{-1} \int d\varepsilon \int_{\varepsilon_k = \varepsilon} \int d\Omega_k f_{k+Q}$$
$$| \int_{-\infty}^{\infty} dt \Delta^{1/2}(t) \exp \left[i\varepsilon t - \int_t^{\infty} Z(\tau) d\tau \right] |^2 \ , \tag{56}$$

with Z given in (12). $\varepsilon_a(z)$ is to include one-body shifts induced by the matrix elements V_k [22]. In (56), f_k is the Fermi function and Q is given by

$$Q = mV \ , \tag{57}$$

where m is mass of the electron and V is the velocity of the projectile parallel to the surface. With $V = 0$, and different initial conditions (because sputtering rather than scattering was being considered), (56) was first found by Nourtier et al., and Bloss and Hone [23]; Brako and Newns applied the ideas to the scattering case [24].

The physics of the two terms n_{a1} and n_{a2} is quite different. The term n_{a1} is called the 'memory' term; it describes the decay of an initial occupation n_0 of the projectile valence level. The physics is most clearly displayed if the projectile level ε_a at infinity lies above the Fermi level, when such a decay will occur if $n_0 = 1$, but not if the projectile orbital is empty ($n_0 = 0$). In the case where ε_a is negative, one may rearrange (55 + 56) to display decay of a hole (ionic) state. The decay of initial memory is a strong effect, because $\Delta(\tau)$ peaks to a large value of order eV at impact with the surface and the integral (55) runs over this time region. Usually decay can be considered to be complete, i.e. memory of the initial state is lost, and $n_{a1} = 0$. An exception is where Δ is of Auger origin, as in He^+ scattering for which ε_a is usually considered to lie below the bottom of the valence band of the target, and then n_{a1} corresponds to the surviving fraction of holes (He^+ions) [25]. The term n_{a2} is related to the charge fraction of particles initially in equilibrium with the surface. In n_{a2}, deviations from adiabaticity occur along the outward section of trajectory. The impact region, where Δ is largest, is not involved. Hence an appropriate approximation is to neglect n_{a1} and to evaluate n_{a2} with a constant outgoing velocity.

We may do this for slow particles not at grazing incidence, in which case if the velocity is much less than the Fermi velocity the parallel velocity effects can be neglected. Then simple formulae were derived by Brako and Newns [24] by applying the method of stationary phase to (56). An example is for the model

$$\Delta(z) = \eta \sin \phi \, e^{-2\alpha z}$$
$$\varepsilon_a(z) = \varepsilon_a - \eta \cos \phi \, e^{-2\alpha z} , \tag{58}$$

where ε_a is the projectile valence level at infinity, η parametrises the magnitude of the surface-induced shift in Δ and ε_a, which is assumed exponential with z, and ϕ is an angle defining the relative magnitude of the shifts in ε_a and Δ.

The model (58) leads to the expression

$$< n_a > = \frac{\sin \phi}{\phi} \exp [- \phi \varepsilon_a / 2\alpha v] , \tag{59}$$

where v is outward velocity normal to surface. We are assuming $\varepsilon_a > 0$ (the opposite case may be obtained by electron-hole inversion). Two obvious limits are, for constant ε_a ($\phi = \pi/2$),

$$< n_a > = \frac{2}{\pi} \exp [- \pi \varepsilon_a / 4\alpha v] \tag{60}$$

a result first found by Nourtier [23], and for $\phi \to 0$,

$$< n_a > = \exp [- \Delta_c / 2\alpha v] , \tag{61}$$

where $\Delta_c = \Delta(z_c)$, and z_c defines where ε_a crosses through the Fermi level. Equation (61), first pointed out by Brako and Newns [24], has been extensively exploited and rederived by Lang and Yu [26]; it has a simple semiclassical interpretation as the survival probability of the filled level after it emerges from the Fermi sea at z_c. Note that the asymptotic results become good approximations at sufficiently low v, when they all have the form $\exp(-C/v)$, as does (59) [24].

A case to which (61) is applicable is the $Li \to Li^-$ process on cesiated tungsten targets [27], in which the affinity level crosses the Fermi level from below on the outgoing path. The data may be analysed assuming $\Delta_c = 0.03$ eV, a remarkably small number. In this case an explanation of the very small value of Δ_c has been provided by Gauyacq [27] in terms of radial correlation within the Li^- orbital. The additional electron sees a $1s^2 2s$ -like core, whose potential V is so weak as to significantly reduce the matrix element $< 2s \mid V \mid k >$.

Notice that correlation effects explain why we have not doubled (59) for spin, which would sometimes be appropriate. Inclusion of the Coulomb interaction U between electrons in the a-orbital in general leads to an upward shift in $\varepsilon_a(z)$, but otherwise no dramatic effects : the problem has been treated by Hartree-Fock [28] and 1/N approximations [29].

In the charge transfer problem we encounter no Infra-Red singularity effects as observed in section 2. Furthermore, introduction of U seems to lead to no appearance of the Kondo energy scale T_K, presumably because charge rather than spin fluctuations are involved [29].

Considering now the effects of parallel velocity, an example where (58-60) are found able to explain the data is the scattering of H^+ (or H^-) off cesiated tungsten surfaces to form H^-. The independence of the initial charge state of the projectile has been explicitly confirmed [20]. It was found possible to explain the data with a fitted function $\varepsilon_a(z)$, strongly supporting (58-60) [20].

We now turn to a rather direct Fermi surface effect, the observation of the internal state of a degenerate projectile valence orbital. Consider neutralisation of a grazing incidence proton into the $n=3$ or $n=2$ shell, both of which lie above the Fermi level e.g. of a Ni target. Due to the image effect which leads to an upward shift in $\varepsilon_a(z)$ approximately given by

$$\varepsilon_a(z) = \varepsilon_a + e^2/4z \quad , \tag{62}$$

these levels in practice lie well above the Fermi level at practical distances from the surface. In grazing incidence scattering experiments the normal velocity v is small and the mechanisms already discussed (e.g. leading to the result (59) with $\phi > \pi/2$) cannot lead to significant occupation of the a-orbital. Occupation of the orbital instead comes from the pseudo-thermal distribution f_{k+Q}. This distribution extends from $\varepsilon_F - k_F Q$ to $\varepsilon_F + k_F Q$, i.e. it involves a 'temperature' of order $k_F Q$; it might be termed the 'Doppler-Fermi-Dirac' distribution. For proton energies of several KeV the DFD distribution can extend above the energies of the $n=2$ and $n=3$ levels of Hydrogen, populating them. A new 'thermal' saddle point of (56) becomes important, leading as we shall see to a statistical-type result.

To sketch this derivation, we start by extending the Anderson model to include projectile orbital degeneracy:

$$\begin{aligned}
H &= \sum_k \varepsilon_k n_k + \varepsilon_a(z) \sum_L n_L \\
&+ \sum_{k,L} V_{kL} u(z)[C_k^+ C_L + \text{h.c.}] \quad .
\end{aligned} \tag{63}$$

In (63), $L = (l,m)$ denotes the state of the nth hydrogen shell of energy $\varepsilon_a(z)$, otherwise the notation is the same as (1-3). A separable form for $V_k(z)$ has been assumed as in (10). We wish to calculate the intrashell density matrix, which in Heisenberg picture is

$$\rho_{LL'} = \; < t_0 \mid C_L^+(\infty) C_{L'}(\infty) \mid t_0 > \quad . \tag{64}$$

Setting up equations of motion, the analogue to (8-9), and eliminating the C_k, we obtain

$$i \frac{dC_L}{dt} = \varepsilon_a(t) C_L - iu(t) \sum_L \int_{t_0}^{t} dt' K_{LL'}(t - t') u(t') \tag{65}$$

$$\times \; C_{L'}(t') + \overset{\wedge}{\phi}_L(t) \quad , \qquad \text{where}$$

$$K_{LL'}(\tau) = \pi^{-1} \int \Gamma_{LL'}(\omega) e^{i\omega\tau} d\omega \quad , \tag{66}$$

$$\Gamma_{LL'}(\omega) = \pi \sum_k V_{Lk} V_{kL'} \delta(\omega - \varepsilon_k) \qquad \text{and} \tag{67}$$

$$\overset{\wedge}{\phi}_L(t) = \sum_k C_k^0 V_{kL} u(t) e^{-i\varepsilon_k(t - t_0)} \quad . \tag{68}$$

The trick is to diagonalise $\Gamma_{LL'}$ for the energies of interest, going to a new basis ij, where

$$\sum_{LL'} R_{iL}^{-1} \Gamma_{LL'} R_{L'j} = \Delta_i \delta_{ij} \quad . \tag{69}$$

The equations of motion (65) may now be separated and solved in the usual approximation to give a generalization of (56)

$$\rho_{ij} = \sum_k W_{ki} W_{jk} f_{k+Q} \int_{-\infty}^{\infty} dt \int_{-\infty}^{\infty} ds \; e^{-i\epsilon_k(t-s)} u(t)u(s)$$

$$\times \exp\left[i \int_s^t \epsilon_a(\tau) - \int_t^\infty \Delta_i(\tau)d\tau - \int_s^\infty \Delta_j(\tau)d\tau \right] \quad , \quad \text{where} \tag{70}$$

$$W_{ik} = \sum_L R_{iL} V_{Lk} \quad , \tag{71}$$

$$\Delta_i(t) = u^2(t)\Delta_i \quad .$$

To approximate (70) by stationary phase, we assume that V is very large, v very small (i.e. trajectory very grazing), so that $\sum_k W_{ki} W_{jk} f_{k+Q}$ is a relatively smooth function of ϵ_k near $\epsilon_a(z)$. The integral (70) is written in terms of $\Delta t = (t - s)$ and $T = (t + s)$, and made stationary w.r.t. Δt and ϵ_k ; the T-integral may be performed exactly for the exponential form $u(t) = \exp(-\alpha vt)$. A necessary condition for the saddle point approximation to be valid is $\alpha v << k_F V$. We then have

$$\rho_{ij} = (\Delta_i + \Delta_j)^{-1} \frac{1}{4} \int d\Omega_k W_{ki} W_{jk} f_{k+Q} \; \rho(\epsilon_k) \Big|_{\epsilon_k = \epsilon_a(z_{ij})} \quad , \tag{72}$$

where ρ is the density of metal electron states and

$$2\alpha z_{ij} = \ln((\Delta_i + \Delta_j)/2\alpha v) \quad . \tag{73}$$

The result (72) shows that ρ_{ij} is a weighted average of the matrix element product $W_i W_j$ over the DFD distribution at energy $\epsilon_a(z_{ij})$, an effective hydrogen valence level energy at a distance given by (73), which turns out to be quite large, of order 10 a.u..

It is straightforward to recalculate $\rho_{LL'}$ from ρ_{ij} and therefrom determine the polarization parameters S/I, C/I, and M/I of light emitted normal to the scattering plane in an $n = 2 \rightarrow n = 1$ or $n = 3 \rightarrow n = 2$ transition. We earlier considered the the n=3 case [4]. In Fig. 4 we illustrate the polarization parameters plotted versus velocity for the n=2 $\rightarrow$ n=1 transition [30], the calculation being done in exactly the same way as in the previous case [4], except that we improved our saddle point determination in the manner just described [31]. Indeed, the calculation was done before data for the n=2 $\rightarrow$ n=1 transition became available. The earlier saddle point method [4] fits the data rather better, but the overall description of the data is satisfactory considering that this is a parameter-free theory. A defect is that the present theory always gives C/I = 0. We believe that the reason for this is the neglect of distortion of the image charge due to its 'lagging behind' the proton in its fast motion along the surface, an effect which breaks the rotational symmetry of the potential about an axis through the proton and normal to the surface, and leads to nonzero C/I.

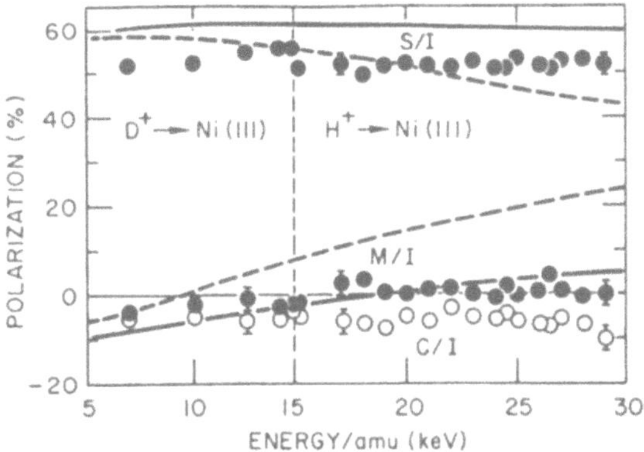

Fig. 4.   Polarization parameters S/I, M/I, and C/I in n=2 → n=1 transition plotted vs. parallel velocity expressed as proton energy.   Full circles, S/I (upper) and M/I (lower), empty circles C/I.   Full curve represents earlier saddle point approximation, broken curve results of (72-73); upper curves S/I, lower curves C/I.

5. Summary

In this article we have outlined some of the results obtained when an Anderson model description of the electronic interaction between an atom and a surface is applied (using the trajectory approximation) to several scattering problems involving electronic degrees of freedom.   What has been learned?

It has been found that the excitation of electronic degrees of freedom is usually not as good a channel for energy transfer in atom-surface scattering as the phonon degrees of freedom.   But, the electron-hole channel should show up as a reduction in the Debye-Waller factor when careful calculations excluding it are compared with experiment.   This may already be apparent in the work of Brenig et al. [11] on NO/ Ag (111) scattering.

The converse situation occurs when tightly bound molecules are scattered non-dissociatively from metal surfaces, the coupling of molecular vibrations to phonon degrees of freedom is weak, and the residual (still small) coupling to electron-hole degrees of freedom dominates the energy exchange between the vibrator and the surface.   This area seems to be especially fertile for study, particularly in areas such as nearly dissociative scattering and the study of intrinsic precursors.

The theory of charge transfer successfully explains the behaviour of charge fractions as $\exp(-C/v)$, where v is normal velocity, but has a tendency to seriously overestimate C relative to experiment.   The case of Li → Li⁻ has been discussed in this paper, and in this case an interpretation of C in terms of the reduction in the lifetime broadening Δ of the 2s-orbital because of radial correlation within the Li⁻ ion has been put forward.   It may be that the explanation is more general, i.e. that electron correlation effects are responsible for the reduction in C, for example via renormalisation of the lifetime broadening, but more work is needed to explore the detailed mechanism in other cases.

A rather direct type of experiment involves determination of the internal state of degenerate valence shells such as H n=2, following pickup of an electron at the surface in grazing incidence collisions.   The theory is relatively free of many-body complications since only one electron is transferred into an empty shell.   The theory of this process involves introducing the concept of the Doppler-Fermi-Dirac distribution whereby pseudo-thermal excitation occurs by relative motion alone.   The theory is rather satisfactory in explaining the existing experimental data.

115

<u>References</u>

1. E. Muller-Hartmann, T.V. Ramakrishnan, G. Toulouse: Phys. Rev. B3. 1102 (1967); Solid State Commun. 9, 99 (1967)

2. C.T. Rettner, F. Fabre, J. Kimman, D.J.Auerbach: Phys. Rev. Lett. 55, 1904 (1985)

3. D.M. Newns: Surf. Sci. 167, 60 (1986)

4. R. Brako, D.M. Newns, N.H.Tolk, J.C.Tully , R. Morris: Phys. Lett. 114A, 327 (1986); R. Brako: Phys. Rev B30, 5629 (1986)

5. K. Schönhammer, O. Gunnarsson: Z. Phys. B38, 127 (1980); Phys. Rev. B 22, 1629 (1980)

6. R. Brako, D.M. Newns: Solid State Commun. 33, 669 (1980); J.Phys.C 14, 3065 (1981)

7. K. Makoshi: J. Phys. C16 3617 (1983)

8. G-P. Brivio, T.B.Grimley: Surf, Sci. 161, L569 (1985)

9. B. Gumhalter, Z. Crljen: Surf. Sci. 139. 231 (1984)

10. G.D. Mahan:<u>Many-Particle Physics</u>. 3rd.ed. (Plenum. New York, London 1981)

11. W. Brenig, H. Kasai, H. Muller: In <u>Dynamical Processes and Ordering at Solid Surfaces</u>.eds. A. Yoshimori and M. Tsukada, Springer Ser.Solid-State Sci., Vol. 59 (Springer, Berlin, Heidelberg 1985). p. 2

12. O. Gunnarson and K. Schonhammer, preprint.

13. B.N.J. Persson, M. Persson: Surf. Sci. 97, 609 (1980)

14. J.A. Barker, private commun.

15. W. Domcke, L.S. Cedarbaum, F. Kaspar: J. Phys. B 12, L59 (1979)

16. B.N.J. Persson, M. Persson: Solid State Commun. 36, 171 (1980)

17. J.W. Gadzuk: J. Chem. Phys. 79, 6341 (1983); 81, 2828 (1984)

18. K.B. Huber, G. Herzberg: <u>Constants of Diatomic Molecules</u> (Van Nostrand - Reinhold, New York, 1979)

19. J.N.M. Van Wunnik, R.Brako, K.Makoshi, D.M. Newns: Surface Sci. 126. 618 (1983)

20. J.N.M, Van Wunnik: thesis, University of Amsterdam (1983)

21. J. Burgdorfer, E. Kupfer, H. Gabriel: Phys. Rev. A35, 4963 (1987)

22. R.Brako, D.M. Newns: unpublished.

23. A. Blandin, A. Nourtier. D. Hone: J. Phys. 37. 365 (1972); A. Nourtier: these d'Etat, Orsay (1972)

24. R.Brako, D.M. Newns: Surf. Sci. 108. 253 (1981)

25. R. Souda, M. Aono, C. Oshima, S.Otani, Y. Ishizawa: Surf. Sci. 172, 657 (1986)

26. M.L. Yu, N.D. Lang: Phys. Rev. Lett. 50. 127 (1983); N.D. Lang: Phys. Rev. B 27, 2019 (1983)

27. J.J.C. Geerlings, R. Rodink, J. Los, J.P. Gauyacq: Surf. Sci., in press.

28. A. Yoshimori, K. Makoshi: Prog. Surf. Sci. 21, 251 (1986); K. Makoshi, H. Kawai, A. Yoshimori: J. Phys. Soc. Japan, 53, 2441 (1984); H. Kawai, K. Makoshi. A. Yoshimori: J. Phys. Soc. Japan 55, 2002 (1986)

29. R.Brako, D.M. Newns: Solid State Commun. 55, 633 (1985);

30. H. Winter: private commun.

31. H. Winter, R. Brako, D.M. Newns: in preparation.

Heavy Fermions

C.M. Varma

AT& T Bell Laboratories, Murray Hill, NJ 07974, USA

These lectures are organized as follows:
1. Introduction
2. Physics Of Occurrence
3. Phenomenology
4. Aspects of the Kondo Problem
5. Problems in Applying Ideas from Kondo Effect to the Heavy-Fermion Lattice
6. Theories Employing Kondo Resonance Plus Bloch's Theorem
7. Two-Impurity Problem
8. The Lattice Problem
9. Correspondence With Experiments

1. Introduction

Professor Kubo in his introductory remarks told us that the Fermi-sea, once thought of as calm as the pacific ocean, has been revealed following Kondo's work of the 1960's to be actually a dangerous sea. I might add that it has also very seductive and mysterious waters. Several generations of theorists have been drawn to it and many have drowned. It is in fact not possible to be a theoretical physicist without concerning oneself with it and the infrared divergences it presents. The heavy Fermion phenomena, on which I have worked for several years is one of the many diverse problems where the physics discovered by Professor Kondo plays an essential role.

2. Physics Of Occurrence

The genesis of the special physics of the rare-earth and the actinide compounds lies in the atomic physics of the f-shell. Because of the large centrifugal repulsion $\ell(\ell+1)/r^2$ with $\ell = 3$ for the f-shell, the 4f orbital is unoccupied, while the 6s, 5p and 5d shells are filled. Consider Ba which contains 2 electrons in the 6s shell outside a Xe core. The 4f orbital has almost zero-binding energy and a large radial extent with a peak at about 14 atomic units from the nucleus (see Fig. 1). Let us now increase the nuclear charge by two arriving at Ce. Because of the incomplete screening of the nuclear charge by the two extra added electrons, all the orbitals 5d, 6s, 6p, etc. are pulled in a little bit. But this effect is small since, for example, the 6s orbital has to be orthogonal to the

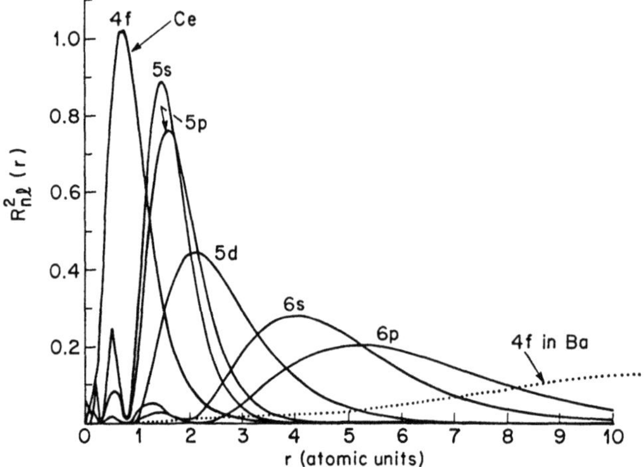

Fig. 1   The charge density of the 4f orbital of Ce in relation to the 4f orbital in Ba and to the charge density in the other orbitals which is almost the same for Ce and Ba.

5s, which has to be orthogonal to the 4s orbital and so on. On the other hand the 4f orbital being the first orbital of its angular symmetry, has no such restriction. The relatively small nuclear charge increase pulls this orbital all the way in so that the maximum in its charge density is at about 0.7 a.u. inside even the core-like 5s and 5p orbitals. But its ionization energy is still small — close to that of the widely spread out 5d, 6s, 6p orbitals. The latter make the wide valence and conduction bands in the solid state. Being core-like the 4f orbitals have large correlation energies — the difference U of the ionization energy I and the affinity energy A for a given charge state in the rare-earths (even after screening effects are taken into account) is large:

$$U = I - A \ . \tag{2.1}$$

This near degeneracy (on the atomic scale of energy) of highly correlated core-like orbitals with the well spread out - weakly correlated-orbitals which form bands in the solid state is at the root of the special physics of the rare-earth and the actinides.

The nature of the phenomena actually observed in the rare-earth and actinide compounds depends very sensitively on the difference between the ionization energy of the f-orbitals and the chemical potential set by the Fermi-energy. The problem is best discussed through the Anderson lattice Hamiltonian first introduced for such problems by Varma and Yafet [1]. This Hamiltonian is depicted pictorially in Fig. 2

$$H = \epsilon_0 \sum_{i\sigma} n_{fi\sigma} + \sum_k \epsilon_k c^\dagger_{k\sigma} c_{k\sigma}$$
$$+ U \sum_{i\sigma} n_{fi\sigma} n_{fi-\sigma} + \sum_k V_k e^{ik \cdot R_i} c^\dagger_{fi\sigma} c_{k\sigma} \ , \tag{2.2}$$

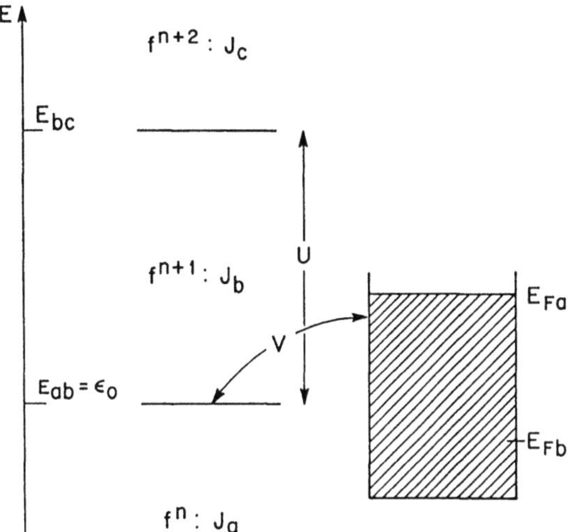

Fig. 2   Pictorial representation of the Anderson model to illustrate the three different regimes of the physics of the rare earths and the actinides.

ϵ_0 is the ionization energy of the f-orbitals which have large correlation energies U and weak hybridization V with conduction orbitals moving in a wide band of dispersion ϵ_k. For understanding the properties of actual materials quantitatively, it is undoubtedly necessary to include the details of multiple orbitals, spin-orbit coupling, crystal fields, etc. But the essential physics is believed already contained in (2.2).

There appear to be three different regimes in the physics of these materials, which are usefully discussed with reference to eqn. (2.2) and Fig. 2. Consider the electronic configurations f^n and f^{n+1} of an isolated ion with respective ground states $|f^n:J_a>$ and $|f^{n+1}:J_b>$. We suppose the ion to be in weak contact with an electron reservoir in which the chemical potential is at energy E_F. There will exist some energy boundary E_{ab} such that at $T = 0$ the ion will be in the $f^n: J_a$ state if $E_F < E_{ab}$ and in the $f^{n+1}:J_b$ state if $E_F > E_{ab}$. Let each neutral atom contain $n + r$ electrons and let us form a crystal out of a large number of them. The non-f electrons go into a set of (mostly) d-bands which constitutes the electron reservoir. Let E_{Fa} be the Fermi level when it contains r electrons per atom and E_{Fb} for $r - 1$ per atom. let there be a weak quantum mechanical contact $\{V\}$ between the two systems,which couples in a tight-binding manner the localized ionic orbitals with the appropriate combination of conduction band orbitals as dictated by the selection rules. Let E_{bc} be another ionization boundary separating the regime of the $f^{n+1}:J_b$ configuration from the $f^{n+2}:J_c$ configuration. $E_{bc} - E_{ab} \equiv U \approx 10$ eV. Figure 2 is of course the pictorial representation of the Anderson lattice Hamiltonian.

a. Magnetic Regime

In this regime, which may be considered understood and occurs in about 90% of the cases, the relevant ionization level of the f-state E_{ab} is far from E_{Fa} or E_{Fb} and therefore the chemical potential is at, say, E_{Fa}. The f-orbitals have no charge fluctuations (integral valence) and may be thought to be in a Mott insulating state. They obey atomic spectral rules supplemented by crystal fields. There is weak residual spin polarization of the conduction electrons, RKKY interactions between the local moments, magnetic transition at low temperatures and spin-wave excitations. The conduction electrons at low temperatures scatter off these spin waves. The degrees of freedom of the local moments are described by spin Hamiltonians and their coupling to conduction electrons can be treated perturbatively.

b. Heavy Fermions

Now suppose we adjust just one parameter so that the ionization boundary, E_{ab}, comes closer to the chemical potential but is still at a separation much larger than $O(\Gamma \approx \pi V^2 \rho)$ where V is the hybridization matrix element (taking into account orbital degeneracy) for f-orbitals with the conducting electrons and ρ the relevant partial density of states of the conduction electrons in the f-channel. In this regime also, at low temperatures, real quantum-mechanical valence fluctuations may be considered negligible. The properties of such materials are radically different from the first regime. Below a certain cross-over temperature T_x, the degrees of freedom of the local moments appear in the form of entropy characteristic of a Fermi-liquid [2]. The same entropy in regime I is of course in the spin waves plus paramagnetic fluctuations above the magnetic transition temperature. Whereas in regime I these are distinguishable and have only a weak coupling between them, in this regime they are so strongly coupled that one cannot speak of them separately. At low temperatures sometimes these degrees of freedom have instability to antiferromagnetic state with weak moments (as in $CeAl_2$, U_2Zn_{17}, etc.) or as superconductivity (as in $CeCu_2Si_2$, UBe_{13}). Some cases are known ($CeAl_3$, $CeCu_6$) where down to the lowest temperature investigated (~ 10 mK) the Fermi-liquid behavior persists.

The specific heat of these materials in the Fermi-liquid regime in terms of the coefficient γ of $(C = \gamma T)$ ranges from about 50 mJ/mole K [2] in $CeAl_2$ to about 1.6 J/mole K [2] for $CeCu_6$. If, as is customary, the specific heat is expressed in terms of an effective mass m^*, $m^* \approx 500$ times the free electron mass for UBe_{13}. Hence the name *heavy fermions*. The magnetic susceptibility at low temperatures tends to a constant as in a Fermi-liquid and its value is enhanced with respect to non-interacting electrons by about the same ratio as the specific heat, i.e. by about m^*/m. Figure 3 reproduces a compilation by Meissner *et al.* [3] of the γ coefficient and the inter-rare-earth distance of several Ce and U compounds and also indicates their known low temperature instabilities.

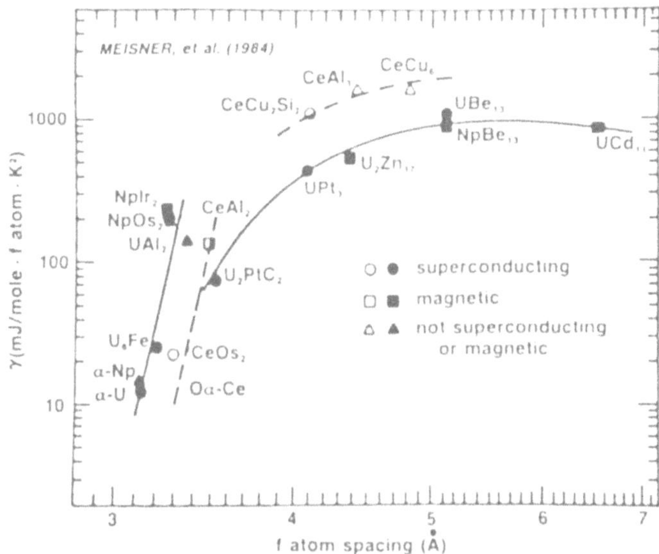

Fig. 3   The coefficient of the linear specific yeat γ for some rare-earth and actinide compounds plotted as a function of the f-atom spacing.

c. Mixed- or Fluctuating Valence Regime

The third regime of the rare earth and actinides [4] occurs if E_{ab} should happen to lie between E_{Fa} and E_{Fb}; then the mixed-valence situation arises (Fig. 2). It requires no remarkable accident of nature for this to happen. The state of the system in which all ions have n f-electrons is impossible because E_{Fa} is too high to be consistent with the f^n: J_a ionic states, and similarly having all ions in $f^{n+1}:J_b$ is impossible because the corresponding E_{Fb} lies below E_{ab}. Thus the Fermi level in the reservoir will lie at the boundary E_{ab} with some fraction x of ions in the f^{n+1} and the rest in the f^n states as demanded by having the correct total number of electrons. This consideration also reveals that about 10% of rare-earth compounds should belong to the intermediate valence category, since typically $E_{Fa} - E_{Fb} \approx 1$ eV while $U \approx 10$ eV.

The third regime occurs frequently in cerium, thulium and ytterbium compounds as well as in samarium and europium. In the former, the underlying close shell makes the $4f$ and the $5d$ atomic energy levels closed, and in the latter two Hund's rule couplings preference for the f^7 configuration makes different configurations close in energy.

The demarcation between regime b and regime c is not always clear from a phenomenological point of view. The various experimental methods of valence determination can only be trusted, at best, to about 10% and the uncertainty in the difference $E_{ab} - E_{Fa}$ is several times Γ. The Fermi-liquid behavior and the high temperature regime of free moments are qualitatively quite similar, although I do not know of any case in a well-established intermediate-valence case that $m^*/m \gtrsim 10^2$, or of a magnetic transition (except in TmSe) or superconducting transition in them. From a theoretical point of view, the intermediate-valence regime must have f-charge and f-moment fluctuations to about equal degree while in regime II real (as opposed to virtual) charge fluctuation may be considered negligible at low temperatures.

In this review, I will deal exclusively with the heavy Fermion regime in which real or resonant charge fluctuations can be neglected. One can then work with a simpler version of the Hamiltonian (2.2) obtained by a canonical transformation [5] to eliminate charge fluctuations:

$$H = \sum_{k} \epsilon_k c_{k\sigma}^{\dagger} c_{k\sigma} \; + \; J \sum_{i} c_i^{\dagger} \, \vec{\sigma} \, c_i \cdot \vec{S_i} \qquad \text{where} \qquad (2.3)$$

$$J \approx \sum_{k} \frac{2\,|V_k|^2\,U}{\epsilon_0(\epsilon_0 + U)} \qquad (2.4)$$

with the Fermi-level taken at zero energy. In (2.2), $\vec{S_i}$ is the local moment at site in the f-orbital i. If there were a local moment only at one site, any $i = 0$, we would have just the Kondo problem characterized by the temperature

$$T_k \approx D(\rho J)^{\frac{1}{2}} \exp(-1/2\rho J) \qquad (2.5)$$

at which the local moment begins to be renormalized. In (2.5), D is the bandwidth of the conduction electrons and ρ the density of states near the Fermi-level.

To second order in J an interaction develops between two moments at sites i and j. This is the well known RKKY interaction

$$H_{int} = \sum_{i<j} K_{ij} \vec{S_i} \cdot \vec{S_j} \qquad (2.6)$$

$$K_{ij} = K_0 f(R_{ij}), \quad K_0 \equiv J^2 \rho \, . \qquad (2.7)$$

For $K_0 \gg T_K$, magnetic order develops and the Kondo effect is moot. The heavy Fermion behavior arises for $T_K > K_0$. This condition can be met in our simplified model only for unrealistically large J of order ρ^{-1}. But for realistic starting points, with orbital degeneracy, crystal field splitting, etc., it is achieved [6] for reasonable values of $J\rho$. Note that the condition $T_K > K_0$ requires, in terms of the Hamiltonian (2.2), that the ionization potential (ϵ_0) not be too far below the Fermi-level and the hybridization parameter V be large enough.

3. <u>Phenomenology</u>

The Hamiltonian (2.3) is suitable for describing the ordinary rare-earth magnetic metals and compounds as well as the heavy fermion compounds. All that is involved is a change in the value of the parameter $J\rho$. In fact at high temperatures the heavy Fermion compounds exhibit magnetic susceptibility of the Curie-form, $\chi \sim \mu^2/T$ with μ nearly equal to the full localized magnetic moment. As the temperature is decreased there is a cross-over to a Fermi-liquid magnetic susceptibility and a concomitant specific heat C_V linearly increasing with temperature. The magnitude of the linear coefficient normalized to the value expected for a Fermi gas defines the effective mass, $C_V/C_{V0} \equiv m^*/m$ is enormously large — of the $O(10^3)$ for several solids. This is the largest renormalization that has ever been observed. Compare this with liquid ^3He where as pressure is increased $m^*/m \approx 6$ near the melting line. It solidifies with further increase of pressure.

I will now give a simple argument for the conditions under which a very heavy fermion mass of $O(10^2-10^3)$ is mandated [7]. Consider the contributions to the entropy of an ordinary rare-earth metal or compound. There is a small linear contribution γT from the fermions corresponding to a mass of $O(1)$, and a contribution from the local moments which at room temperature is nearly temperature independent and two to three orders of magnitude larger than the γT contribution. The spin contribution to the entropy decreases as temperature is decreased, and as illustrated in Fig. 4 has a change in slope at the magnetic ordering temperature T_N which is typically of $O(10K)$. Below T_N, one gets the spin-wave contribution to the entropy. The heavy fermions are described by the same Hamiltonian as the typical rare-earth solid albeit with somewhat different parameters. Suppose that we demand that there be no magnetic phase transition[8] and that the low-energy excitations in these solids be purely fermionic, i.e., there is a gap of $\sim T_F$ which is the same order of magnitude as T_N for spin-fluctuation excitations, then as illustrated in Fig. 4, a heavy-fermion mass of $O(10^2-10^3)$ is necessary, since total entropy in the present problem is the same as for the ordinary rare earths and the high-temperature entropies must be identical.

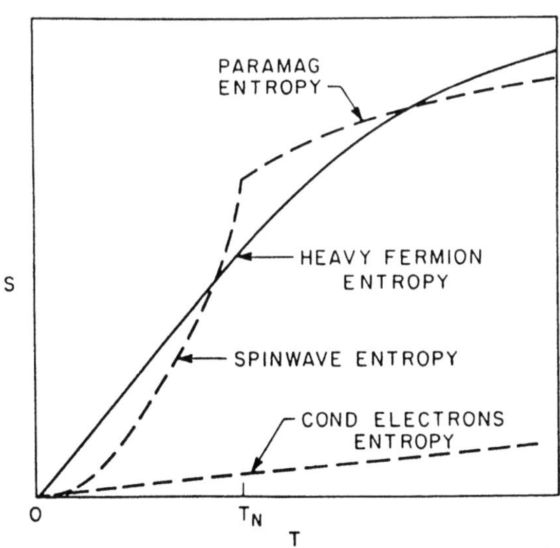

Fig. 4   The magnetic and the conduction electron contribution to the entropy of an ordinary rare-earth magnetic metal (dashed lines) compared with the entropy for a heavy fermion material.

The discussion above may be summarized by the statement that heavy fermions arise from ordinary conduction electrons by exchange of spin fluctuations with total spectral weight similar to that carried by the local moments and that if these spin fluctuations must have a gap of $O(T_F)$, the specific heat of the fermions will be $O(T/T_F)$. One really observes only the spin-fluctuation entropy but in a fermionic form, i.e., $\sim T$.

This point of view is reinforced and at the same time made more precise by the observation that although the thermodynamic properties like the specific heat and the magnetic susceptibility are severely renormalized, many transport properties: the coefficient of this leading low temperature terms in the thermal conductivity, ultrasonic attenuation, nuclear and impurity spin-lattice relaxation rate, the residual resistivity etc. appear unrenormalized [7]. Naively all these quantities depend on $(m^*/m)^2$ and one would expect enormous renormalizations; none are seen. Such a behavior is understood by asserting that in the single particle fermion Green's function

$$G(\mathbf{k},\omega) = (\omega - \epsilon(\mathbf{k}) - \Sigma(\mathbf{k},\omega))^{-1} , \tag{3.1}$$

where $\epsilon(\mathbf{k})$ is the dispersion obtained from some self-consistent one-electron theory, $\Sigma(\mathbf{k},\omega)$ the self-energy function is required to have the property

$$\partial\Sigma/\partial\omega >> \mathbf{v}_0^{-1} \cdot \partial\Sigma/\partial\mathbf{k} . \tag{3.2}$$

$\mathbf{v}_0$ is defined by $\epsilon(\mathbf{k}) = \mathbf{v}_0 \cdot (\mathbf{k} - \mathbf{k}_F)$ near the Fermi-surface. Then near the Fermi-surface

$$G(\mathbf{k},\omega) \approx z(\omega - z\,\mathbf{v}_0 \cdot (\mathbf{k} - \mathbf{k}_F))^{-1}, \tag{3.3}$$

where z the quasi-particle renormalization amplitude is

$$z = (1 - \partial\Sigma/\partial\omega)^{-1} . \tag{3.4}$$

The coefficient of the linear specific heat is then proportional to the renormalized density of states $\sim z^{-1} \int d\mathbf{S_k} |v_0(\mathbf{k})|^{-1}$. Therefore

$$m^*/m \approx z^{-1} . \tag{3.5}$$

On the other hand, with the condition (3.2), the self-energy and the vertex renormalization cancel in most transport properties. This is usually shown through the use of a Ward identity. One may motivate it physically by noting that most transport properties involve a quantity τ^*/m^*, where τ^* is the renormalized scattering time. Now in the Born approximation

$$(\tau^*)^{-1} \sim |<\psi_i^*|M|\psi_f^*>|^2 n_f^*(E) , \tag{3.6}$$

where ψ_i^* and ψ_f^* are quasiparticle wavefunctions, M is the appropriate operator for the transport process under consideration and n_f^* is the renormalized final density of states. If the coherent part of the Green's function, which is proportional to the product of two quasiparticle wavefunctions, is proportional to z, each quasiparticle wave function has a weight $z^{1/2}$. Further $m^* \sim n_f \sim z^{-1}$. We then see that τ^*/m^* is independent of the renormalizations.

These arguments as well as the more formal proof using the Ward-identity work for strong scattering as well. They also survive into the superconducting state provided only that the condition (3.2) is met.

The condition (3.2) is equivalent to renormalizations through fluctuations whose energy scale is much smaller than the Fermi-energy of conduction electrons but whose momentum scale is comparable to the Fermi-wave-vector. Physically the cancellation of the renormalization in transport properties arises because the condition (3.2) is equivalent to the statement that the self-energy rides with the local chemical potential so that in any perturbation where the latter is altered and relaxes towards equilibrium, the renormalizations are always the same and in derivatives with respect to time are not felt.

This situation is very familiar from the case of electron-phonon interactions. Quite apart from the two orders of magnitude difference in the magnitude between heavy fermions and the electron-phonon problem, there is another important difference: Magnetic susceptibility is not renormalized in the electron-phonon problem. For the renormalization of susceptibility as in the heavy fermions, the fluctuations exchanged by the conduction electrons must carry spin.

In Table I, we contrast the renormalizations in heavy fermions with those in liquid ^{3}He in terms of the Landau parameters.

	^{3}He	Heavy fermions
Compressibility $$\frac{dn}{d\mu} = \frac{m^*/m}{1+F_0^s}$$	Renormalized	Almost unrenormalized
Entropic mass m^*/m	$1 + F_1^s/3$	$1 + F_0^s$
Dynamic mass m_d	m	$\approx m^*$
Magnetic Susceptibility χ	$\dfrac{m^*/m}{1 + F_0^a}$	$\dfrac{m^*/m}{1 + F_0^a}$
Transport properties	Usually renormalized	Usually unrenormalized

Table I. Comparison of renormalizations in the heavy fermion problem with those in liquid ^{3}He.

The result $m^*/m = 1 + F_1^s/3$ is derived in one-component Fermi-liquids solely using Galilean invariance. In a two-component situation (we will soon discuss

that microscopically that is how we have to start thinking of the heavy
fermions) in which only one compound gives a linear specific heat and the
effective mass is defined only in relation to it but the momentum is shared
between the two components, Galilean invariance cannot be used. This is true
even if there is no lattice and both the components are continuums.

The dynamical mass m_d in Table I is defined through the operator relation

$$\vec{p} = \vec{J}/m_d \ . \tag{3.7}$$

The result that $m_d \approx m^*$ has important consequence in the superconducting
state where it is measured through the London penetration depth.

It is important to note that the condition (3.2) implies that the shape of the
Fermi-surface is not renormalized by the many-body effects. This is consistent
with the de Haas-van Alphen measurements.

The characteristic energy scales of renormalizations in the heavy Fermions
are characteristic of those in the Kondo effect. If we simply assert that all the
magnetic moments of the periodic array of rare-earth atoms in the heavy
fermion solids undergo a Kondo effect and renormalize the mass of the
conduction electrons interacting with them, the specific heat predicted per
rare-earth atoms at low temperature would be $O(T/T_K)$. If we take T_K as
deduced for similar isolated magnetic impurities in conduction electrons, the
experimentally observed magnitude is obtained.

This phenomenological statement raises a number of important questions
which we shall touch on. Before we do so, an interesting relation between the
single impurity Kondo effect resistivity and the resistivity commonly observed
in the heavy fermion solids is discussed.

Fig. 5 shows the resistivity for a single impurity, the high temperature
$ln(T/T_K)$ dependence, and the asymptotic form $\rho_0[1-(T/T_K)^2]$, where ρ_0 is

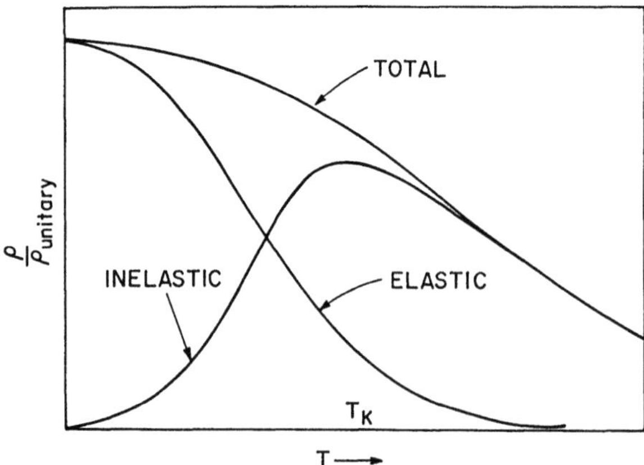

Fig. 5   Decomposition of the total resistivity of a Kondo impurity into an
elastic and an inelastic part.

the unitarity value for an impurity. This resistivity can be strictly separated [9] into the inelastic part, which contributes all of the logarithmic term at high temperatures and which goes to zero as $(T/T_K)^2$ at low temperatures, and the elastic part. In a periodic array of scatterers the elastic part simply leads to a new electronic band and only the inelastic part can lead to electrical resistance. It is indeed remarkable that the resistivity of heavy Fermions is qualitatively similar to the *inelastic part* of the single Kondo impurity. The overall magnitude is such that if the high temperature part is extrapolated to $T = 0$, a value approximately equal to the unitarity scattering from each magnetic ion is obtained. This is another phenomenological argument for believing that each rare-earth moment undergoes some sort of Kondo-effect renormalization.

Many experiments in heavy fermions yield results which are in apparent conflict with those expected of a Fermi liquid. One such is inelastic neutron scattering which yields that the width in frequency of the spectra approaches a constant as the momentum transfer tends to zero. For a Fermi-liquid like ^3He it should go to zero. The general theorem for a Fermi-liquid interacting with other types of excitations is that the response function is given by the (Fermi) quasiparticles if the coupling is to a quantity which is separately conserved by the Fermions in the problem. In heavy fermions we have both the fermions (the conduction electrons) and the local moments which are not fermions. They interact with each other via the exchange coupling and this renormalizes the mass of the former. But because of the exchange interactions the moment of the fermions is not separately conserved. In the neutron scattering experiment therefore there is a large contribution from the local moment which has the q-dependence of the atomic f-form factor. This point has also been made by Pethick and Pines [10].

4. <u>Aspects of the Kondo Problem</u>

Professor Kondo discovered that the perturbation theory in J for the problem of conduction electrons exchange-scattering off a local moment diverges for antiferromagnetic J. The second-order contribution to the T-matrix (see Fig. 6) consists of two terms — one with an intermediate electron line and the other with an intermediate hole line. The quantum mechanical nature of the spins is essential here; the two processes add because of the anti-commutative nature of the spin operators. It was soon realized that the problem cannot be solved by perturbation theory. An important development towards the eventual solution of the problem was the work of Anderson and collaborators [11] who cast it in a

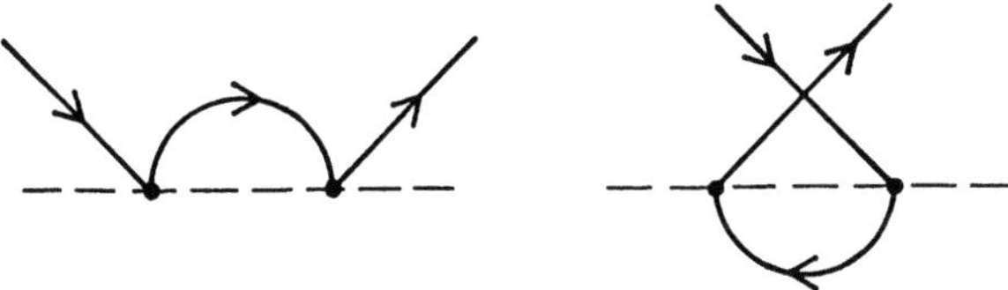

Fig. 6   The leading contributions to the T-matrix for conduction electron-magnetic impurity scattering.

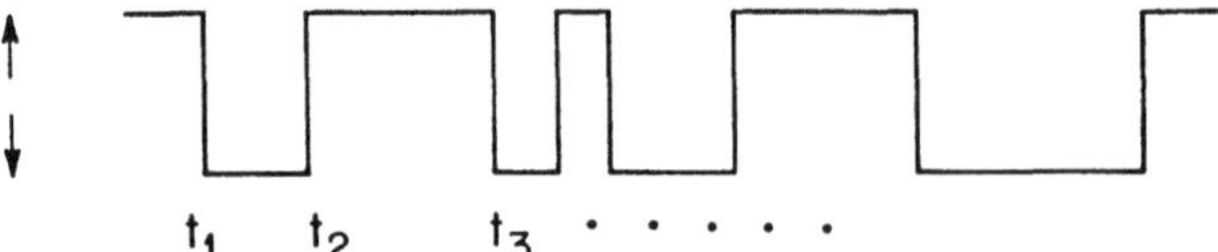

Fig. 7   The path-integral representation of the single Kondo impurity problem.

path-integral representation eliminating the local moment fluctuations. The conduction electron spins scatters up and down (see Fig. 7) at a succession time t_1, t_2, ... The solution of the problem consists in solving for electrons near the Fermi-sea moving in such a time dependent potential. One conclusion of their work is that T_K is only a cross-over temperature and a singlet state of the local moment and conduction electrons is arrived at only at $T = 0$. Through the work of various people it was realized qualitatively that at low temperatures the Kondo problem is equivalent to a resonance at the Fermi-energy of width T_K.

The complete and enlightening solution was achieved by Wilson [12] with a numerical renormalization group method. An analytic solution was obtained by Andrei [13] and by Wiegmann [14]. Wilson's solution was interpreted in a phenomenological fashion by Nozieres [9]. For our subsequent discussion on heavy Fermions it is necessary to review these. For the same reason we review first the important work by Yamada and Yosida [15].

Yamada and Yosida's work is an interesting application of the principle of analyticity and continuity which underlies the derivation for instance of Fermi-liquid theory. Consider the Hartree-Fock solution of the Anderson model for the single magnetic impurity problem. In general there are two solutions: (i) A magnetic solution (Fig. 8a) with a local resonance for a given spin direction below the Fermi-energy direction above the Fermi-energy and a local resonance for the opposite spin. Each resonance has a width $\sim V^2 \rho$ and the two are separated by U. (ii) A non-magnetic solution (Fig. 8b) with a half-occupied resonance also of width $\sim V^2 \rho$ straddling the Fermi-energy. For the range of parameters of interest the magnetic solution has the lower energy in the Hartree-Fock approximation. But we know that the correct ground state is non-magnetic. Most of the brilliant (and diffcult) methods used in the Kondo problem start from the broken symmetry magnetic solution and work hard to restore symmetry and obtain the magnetic solution. Yamada and Yosida's point is that even though U may be a large parameter, it is legitimate to calculate in perturbation theory *about the non-magnetic* Hartree-Fock solution. The solution will analytically and continuously give several features of the right answers if the starting point has the same symmetry as the correct solution. The low temperature "Kondo resonance" is in this picture just a modification in width and strength of the Hartree-Fock non-magnetic resonance. Yamada and Yosida were able to show that at each order in perturbation theory in U/V, the Wilson ratio — the dimensionless ratio of the magnetic susceptibility to the coefficient of the linear term in the specific heat is 2. It is a reward for adopting the correct point of view that the absolute answers for the magnetic

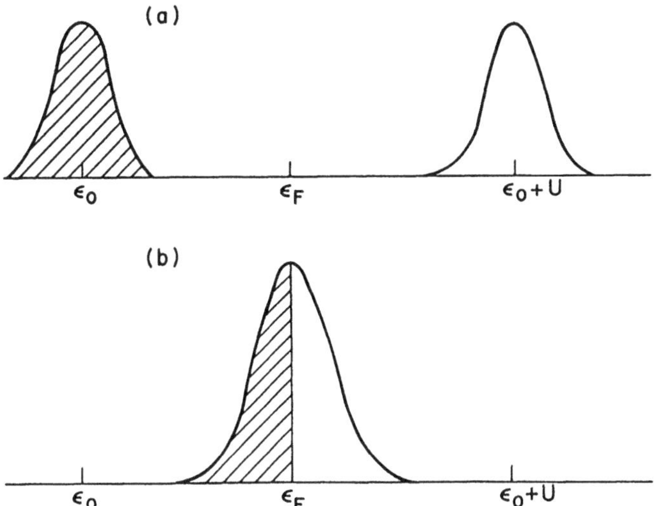

Fig. 8   The Hartree-Fock resonances in the Anderson model for a magnetic impurity. (a) Resonances in the magnetic solution. (b) Resonance in the non-magnetic solution.

susceptibility to within a few percent are obtained already in fourth-order perturbation theory in U/V which may be as large as 10^2 [16]!

Certain aspects of heavy Fermions can be obtained from just such a principle of analyticity. As mentioned, the shape of the Fermi-surface is given by self-consistent one-electron theory provided one treats the f-states as propagating states. Also the Fermi-surface for rare-earth magnetic metals is correctly given by self-consistent one-electron theory provided one treats the f-states as local core-like states with frozen charges. The analyticity principle sometimes does not go too far. For instance one may have certain global symmetries which are the same as those in a simpler problem but in addition have some additional local constraints which are not present in the simpler problem. The additional constraints may then not appear by doing perturbation theory about the simpler problem. For example, we will see in the problem of two Kondo impurities an asymptotic Fermi-liquid behavior but with a correlation between the magnetic moments. So far it has not been possible to see such correlations emerging in perturbation theory about two non-magnetic Hartree-Fock resonances.

d. Wilson's Method for the Kondo Problem

Wilson's Renormalization Group generates a sequence of Hamiltonians suitable for describing accurately the properties of successively lower and lower temperatures. Consider the Kondo problem

$$H_K = J \vec{S}_0 \cdot \psi_0^\dagger \sigma \psi_0 + t \sum_i \psi_{i\sigma}^\dagger \psi_{i+n} \ . \tag{4.1}$$

The effective Hamiltonians at intermediate temperatures are very complicated indeed but for $T \to 0$, they approach the fixed point Hamiltonian

$$H^* = t \sum_{i \neq 0} \psi_{i\sigma}^\dagger \psi_{i+n} , \qquad (4.2)$$

where the impurity is simply blocked out and one electron is effectively lost (it becomes incoherent) to the quasi-particles. This may in perturbation theory language be said to have happened because an effective $J \to \infty$ as $T \to 0$; this eliminates the local moment and an electron out of the problem. The asymptotic low temperature properties are obtained by adding to H^* the leading irrelevant operators about it:

$$\begin{aligned} H(\text{low } T) = H^* &+ \tilde{t}(\psi_0^\dagger \psi_1 + c.c.) \\ &+ \tilde{U} \, n_{0\sigma} n_{0-\sigma} \end{aligned} \qquad (4.3)$$

It is found that $\tilde{U} = \tilde{t} \approx T_K^0 \approx 3 T_K$. $\tilde{t}$ then leads to a resonance of width T_K^0 at the Fermi-energy and $\tilde{U}$ leads to quasi-particle interactions induced by the Kondo effect.

Nozieres observed that (4.3) is equivalent to scattering from a local potential with phase-shift:

$$\delta_\sigma(\epsilon) = \pi/2 + 0.3 \, \epsilon/T_K + \sum_\sigma \phi_{\sigma\sigma'} \delta n_{\sigma'} \qquad (4.4)$$

The Fermi-liquid term $\phi_{\sigma\sigma'}$ was directly related to T_K by assuming the Kondo resonance moves rigidly with the chemical potential. This guarantees $\tilde{U} = \tilde{t}$ in (4.3).

5. Problems in Applying Ideas from Kondo Effect to the Heavy—Fermion

There are two principal questions which are raised which we shall now discuss. The magnetic moments interact (via the conduction electrons) with each other through RKKY interaction. How is then one to talk about the Kondo effect of the individual moments. It used to be argued that if T_K is high compared to K_0, the characteristic RKKY energy, the effect of the RKKY interactions may be neglected because below T_K the moments are locally compensated to singlets and such singlets do not interact with each other! This point of view is incorrect. The interaction between the moments modifies the Kondo effect in an interesting and important way which we shall soon discuss.

The other question arises from the idea, often discussed, of a "Kondo length" or the related idea of the "exhaustion principle" which confronts us if we believe that an electron in an energy scale within T_K of the Fermi-energy is needed to compensate a moment. The concept of a Kondo length arises if we think of the Kondo effect as a bound state with energy T_K; then the spatial extent of the bound state is the "Kondo length" v_F/T_K. Actually there is no experiment nor any calculation [17] which shows the relevance of the concept of a Kondo length. I seriously doubt that if the ground state correlation function $<S. \, \sigma(r)>$ is calculated any length scale besides k_F^{-1} will make its appearance. It is possible that for finite frequency (of the order of T_K) properties, such a length scale appears. The point may well be that the Kondo effect is not a bound state problem but a problem of resonance. The total phase-shift is simply fixed by Friedel's sum rule to be π and the ground state correlations show merely the Friedel oscillations. It is also well to remember the compensation theorem [18], which states that no magnetic moment is ever induced in the conduction electrons.

On the other side there is a considerable body of phenomenological argument that every moment is compensated in the heavy fermions. I think it is best to think that a given conduction electron state interacts with the local moment in each unit cell; the fact that the incoming electron state at a given unit cell arises from the interference due to similar scattering at the other cells merely changes the magnitudes of the parameters $V_\ell(\mathbf{k})$. This multiple interference of course gives rise to new bands in angular momentum channels of the local moments.

6. <u>Theories Employing Kondo Resonance Plus Bloch's Theorem</u>

Most of the approaches to the heavy fermion problem combine in one fashion or the other the known solution to the single impurity Kondo problem with the requirement that the solution to a lattice problem obey Bloch symmetry. For an isolated impurity the Kondo problem is equivalent at low temperatures to the scattering of conduction electrons with the same local symmetry as that of the localized orbital with a phase shift given by eq. (4.4). A local potential $V(r - R_i)$ can now be found to simulate the first two terms. If now the lattice problem is regarded as scattering of conduction electrons by a periodic potential

$$V_{lattice}(r) = \sum_i V(r - R_i) \,,$$

the solution is a hybridized band with width of order T_K as in fig. 9 . Slightly more complicated situation arises if orbital degeneracy etc. is considered.

The philosophy is implemented in one way or other in all of the following schemes:
  i.   Green's function decoupling [1].
  ii. 1/N expansions [19].
  iii. Self-consistent Green's function schemes using Kondo vertex [20].
  iv. Gutzwiller variational schemes [21], and most directly
  v.   phase shift from Kondo plus band structure calculations [4].
The Green's function decoupling schemes were the first to get results of the form discussed above in the mixed-valence context simply by writing down equations of motion and decoupling them in terms involving U which tend to block the motion of local electrons. These results are equivalent to dropping the term U in (2.2) by renormalizing the hybridization; i.e. an effective Hamiltonian

$$H_{eff} = \sum_k \epsilon_k c_{k\sigma}^+ c_{k\sigma} + \sum_i \tilde{\epsilon}_0 c_{i\sigma}^+ a_{i\sigma} + \sum_k (\tilde{V}_{k\sigma} c_{k\sigma}^+ a_{i\sigma} + c.c) \qquad \text{with} \qquad (6.1)$$

$$\tilde{V}_{k\sigma} = <1 - n_{-i\sigma}>^{\frac{1}{2}} V_k \qquad (6.2)$$
$$\tilde{\epsilon}_0 = \epsilon_0 + U <n_{-\sigma}> \,. \qquad (6.3)$$

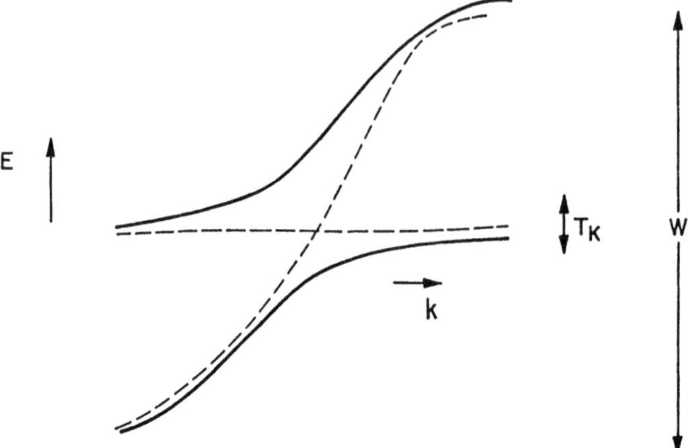

Fig. 9   The band-structure in the pseudo-one-electron solutions of the heavy fermion problem.

The physics simply is that it takes two powers of V for a correlated particle to go from one site i to another site j. For U very large it can do so only if the site j is unoccupied. In mean field theory this hopping reduction is counted as in eq. (6.2).

The connection to the Kondo problem arises from the fact that in the Anderson model

$$\tilde{V}_k^2 \rho \approx T_K \, , \tag{6.4}$$

where $\rho \approx 1/W$ is the density of states of conduction electrons.

The Gutzwiller variational method has exactly the same effective Hamiltonian; eq. (6.1). For large degeneracy, the blocking factor in eq. (6.2) is modified as might be expected; this has the consequence of altering conditions for the stability of the pseudo-one-electron-type solution, eq. (6.3).

One nice point illustrated by the variational solution is the reason for the stability of the simple pseudo-one-electron-type solution. As U is increased, the effective one-electron level moves up, eq. (6.3), depleting the number of electron in the localized orbitals and increasing the occupation of conduction electron states *above the Fermi-energy*. This is the *incoherent part* of $<n_k>$, which has a perfectly good discontinuity at the Fermi-level consistent with Luttinger's theorem. Keeping U fixed if one decreases ϵ_0 below the Fermi-level, $<n_f>$ increases as the cost of transferring electrons to the uncorrelated band increases. Ultimately $<n_f> = 1$, so that $V = 0$ marking a transition to localized (magnetic) state of the correlated orbitals and a new Fermi-surface for the conduction electrons.

The pseudo-one-electron theories of the type discussed above have many virtues and a few serious shortcomings. They are based on the correct symmetry — that of a Fermi-liquid, and therefore give various correct qualitative features of the experiments like the Fermi-surface; they have the correct scale of $d\Sigma/d\omega$, therefore the right order of magnitude of the specific heat and the susceptibility (although not the correct ratio between them). $d\Sigma/dk = 0$ in such theories — so they give the correct answers for the compressibility and the cancellations of renormalization in the transport properties as discussed earlier. The qualitative temperature dependence of the resistivity and its relationship to the specific heat is also explained.

The shortcomings have to do with the fact that they do not attempt the question of interactions between the Kondo resonances and the interplay of such interactions with the scattering of conduction electrons. They dispense with the local moments right away and with that the RKKY interaction between the moments. As we will see below the effect of such interactions cannot be treated in perturbation theory about the single impurity type solutions. The magnetic correlations and the excitation spectra exhibited by them are based on band-structures of the type shown in Fig. 9 . The magnetic correlation are then of the Lindhard or RPA form:

$$\sum_h \frac{f_{k+q}-f_k}{\omega+\epsilon_{k+q}-\epsilon_k}\, g(k,q)$$

with ϵ_k taken from the "band structure" and $g(k,q)$ the appropriate matrix elements to take into account the variation of the orbital character in the bands with k. Since the excitation spectra and the quasiparticle interactions are incorrectly given by such theories, they predict incorrectly the instabilities — antiferromagnetic, superconducting etc. — of the heavy Fermi liquid.

7. <u>Two-Impurity Problem</u>

In order to study the influence of the interactions among the local moments on their renormalization by the conduction electrons, we have solved the two Kondo impurity problem [23] by Wilson's method. Actually we take a somewhat simplified version of the two-impurity problem — but one which, we believe, contains all the essential physics. Besides shedding light on the heavy fermion problem, the two-impurity problem is very interesting in its own right. It belongs to the class of problems in which backward scattering is crucial and which are not soluble by known analytic methods. Earlier, we talked briefly of the path integral formulation of the single impurity Kondo problem. A similar formulation is possible for the two-impurity problem. In this, (see Fig. 10) conduction electrons flip at times t_{1A}, t_{2A}, ... corresponding to impurity spin at site A and at t_{1B}, t_{2B}, ... corresponding to impurity at site B. The new and interesting physics arises because the sets of times $\{ti_A\}$ and $\{ti_B\}$ are self-consistently correlated. This implies that a phase relationship can develop between the two moments even while they dynamically respond to the conduction electrons. Indeed the numerical renormalization method solution discussed below shows such a phase relationship.

For two impurities it is best to work in a basis set of even and odd parity about the mid-plane between the two impurities. In terms of operators, $c_{k\epsilon}$,

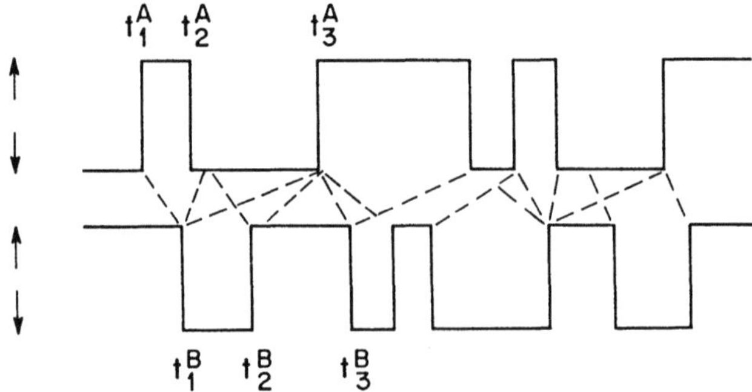

Fig. 10 Path integral representation of the two Kondo impurity problem illustrating the correlations between the times of spin flips of conduction electrons at the two impurities.

c_{k0} for even and odd states and with the simplification of confining the magnitude of **k** to the Fermi-surface, the two impurity problem is equivalent to

$$H = \sum_{k,k'} (\vec{S}_1 + \vec{S}_2) \cdot (J_e c^+_{k'e} \vec{\sigma} c_{ke} + J_0 c^+_{k'0} \vec{\sigma} c_{k0})$$
$$+ (\vec{S}_1 - \vec{S}_2)(iJ_m c^+_{k'e} \vec{\sigma} c_{k0} + h.c.) + H_e + H_0 , \qquad (7.1)$$

where J_e and J_0 are the coupling constants for the even and the odd channels, J_m mixes the two channels and scatters from the singlet to the triplet state of the two impurities and H_e and H_0 are the kinetic energy operators for the even and odd channel electrons. In this model the magnetic interaction between the two impurities is given by

$$K S_1 \cdot S_2 \text{ where } K = (J_e^2 + J_0^2 = 2J_m^2) . \qquad (7.2)$$

Just as for the Kondo problem the asymptotic low temperature behavior for the two-impurity problem is expressible by an effective Hamiltonian consisting of a fixed point Hamiltonian H^*, a one-electron Hamiltonian $H^{(1)}$ and an interaction Hamiltonian H^{int}:

$$H_{eff}(12) = H^* + H^{(1)} + H^{int} . \qquad (7.3)$$

We first discuss H^*, which describes the ground state. In the ground state we lose an electron both in the even parity and the odd parity charge. This is in itself a remarkable result which is quite contrary to what one gets from considering the problem as a "sum" of two one-impurity Kondo problems. In this case the scattering from the left-impurity interferes with the scattering from the right-impurity such that an even parity combination of states moves down in energy with respect to the odd parity combination of states. Integrated up to the chemical potential, the even parity resonance would then have more than one electron and the odd parity channel less than one electron. The sum in the two channels is fixed by Friedel sum rule to be two. Our result that there is one electron in each channel is therefore very interesting. This means the two-impurity problem cannot be represented by local potentials at the two sites. Since one electron in a channel implies the moment in channel is renormalized to zero, the implication is that it is not meaningful to speak of renormalization of individual moments. It is the eigenstates — even and odd parity — where the total moment is zero.

From a study of the deviation from the fixed point, we conclude that the ground state expectation value

$$<S_1 \cdot S_2> \neq 0 . \tag{7.4}$$

Now if we express a fixed-point Hamiltonian (describing $T = 0$ properties) with $J_1 = J_2 = \infty$, the only way we can get $< \vec{S}_1 \cdot \vec{S}_2 > \neq 0$ is if the fixed point Hamiltonian contains a term $K \vec{S}_1 \cdot \vec{S}_2$ with $K = \infty$ with fixed ratios J_e/K, J_0/K. Since we deduce this only from the deviation from the fixed point and not from the asymptotic $T = 0$ spectra, this may not be the technically pure way of expression. Semantics aside we have the result $<S_1 \cdot S_2> \neq 0$ although the total moment, local moments plus that of conduction electron, is zero. We therefore have a correlated Kondo effect.

One-electron Hamiltonian expresses just the hopping on and off the even and odd combination of the two impurity orbitals.

$$H^{(1)} = t_e(f_{0e}^+ f_{1e} + c.c) + t_o(f_{0o}^+ f_{1o} + c.c). \tag{7.5}$$

The important point here is that t_e and t_o increase as the initial Rudermann-Kittel interaction increases. In other words the width of the resonance is not the single-impurity Kondo temperatures.

This result is a direct consequence of the correlation among the local moments, $<S_1 \cdot S_2> \neq 0$ which induces similar ferro- or antiferromagnetic correlations between the quasiparticles at the two sites as discussed below and renormalizes the quasiparticle resonance widths.

While the single-impurity Kondo effect has just one interaction term, the two-impurity problem has in general five quasiparticle interaction terms:

$$\begin{aligned}
H^{int} = & + \sum_p U_p(n_{0p} - 1)^2 \\
& + U_{e0}\{(n_{0e}-1)(n_{0a}-1) + \sum_a (f_{0e\alpha}^+ f_{0e-\alpha}^+ f_{0a-\alpha} f_{0a\alpha} + c.c)\} \\
& + J_{e0} f_0^+ \vec{\sigma} f_{0e} \cdot f_{0o}^+ \vec{\sigma} f_{0o} ,
\end{aligned} \tag{7.6}$$

where $\alpha = \uparrow, \downarrow$ and n_{0p} is 0, 1, and 2. The first two terms are the intrachannel direct interaction, the third is the interchannel direct interaction, the fourth the pair hopping from one channel to the other and the fifth the interchannel exchange interaction. For the simplified two-impurity model solved by us in which the interaction among the two channels is effectively a delta function in space, the third and the fourth terms have equal coefficients.

It is instructive to re-express H_{int} in terms of operators at the two sites i and j. This leads to the following terms

$$\sum_\sigma \mathbf{U}^{(1)} n_{i\sigma} n_{i-\sigma} + i \rightarrow j \quad : \quad \text{on-site repulsion} \tag{7.7}$$

$$\mathbf{U}^{(2)} n_i n_j \quad : \quad \text{intersite repulsion} \tag{7.8}$$

$$\mathbf{J}\, \vec{S}_i \cdot \vec{S}_j \quad : \quad \text{intersite exchange} \tag{7.9}$$

$$\mathbf{P}(c_{i\uparrow}^{+} c_{i\downarrow}^{+} c_{j\downarrow} c_{j\uparrow} + h.c.) \quad : \quad \text{pair hopping} \tag{7.10}$$

$$\mathbf{B}(1 - n_{i\sigma}) c_{i-\sigma}^{+} c_{j-\sigma} \quad : \quad \text{kinetic energy blocking,} \tag{7.11}$$

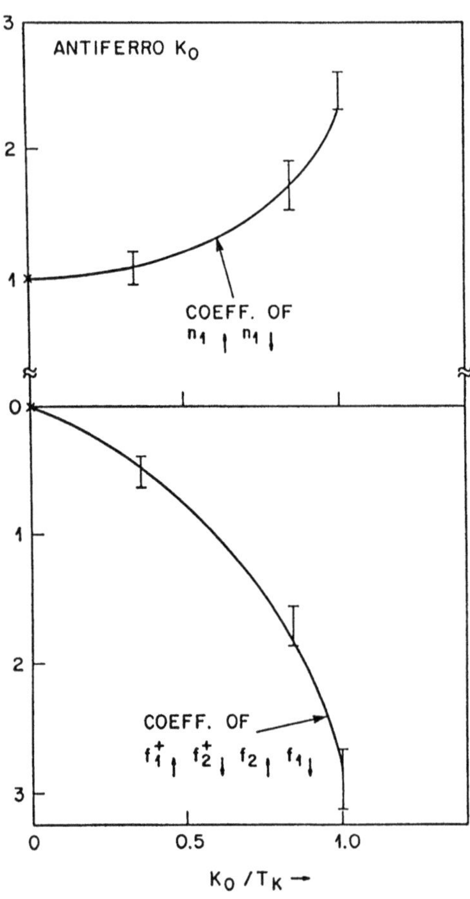

Fig. 11 Coefficients of the on-site repulsion and the inter-site exchange terms in the effective low temperature Hamiltonian of the two impurity problem as a function of the ratio of the initial RKKY interaction between the two to the single-impurity Kondo temperature.

136

The coefficients of all five terms as well as the kinetic energy operator on J and K. For $K = 0$, the isolated impurity problem, only the term (7.7) survives. The coefficients in (7.7) - (7.11) are linearly related to those in (7.6). In Fig. 11 J and $U^{(1)}$ are plotted against K/T_K.

From Eqs. (7.7-7.11), we can calculate the direct susceptibility X_0 and the staggered susceptibility X_s. The results are shown in Fig. 12 . For AFM interactions X_0 departs from the Wilson value as K is increased and X_s increases from zero. There is no universality in the problem and Wilson ratio is not a useful quantity. For very large K/T_K, the two moments form a singlet and no Kondo effect occurs. In between the small K/T_K and the large K/T_K is a point at which the quasiparticle staggered susceptibility diverges at $T \rightarrow 0$ and the Kondo renormalization of the two moments also occurs.

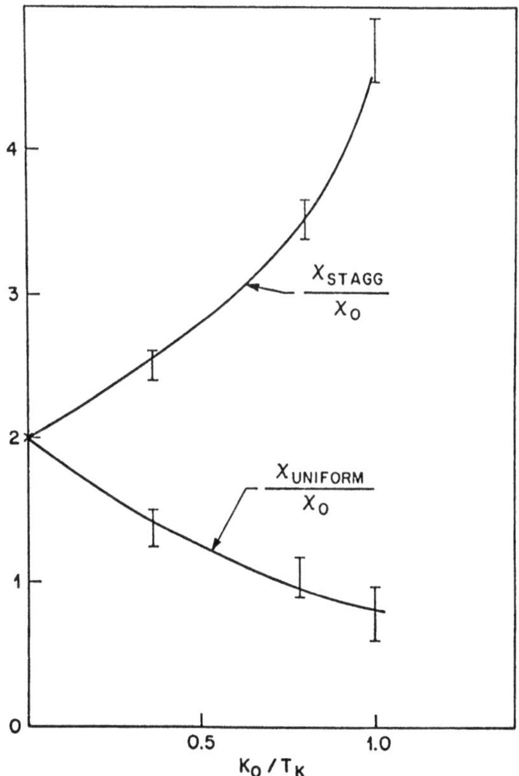

Fig. 12   The uniform and staggered susptibilities in the two Kondo impurity problem.

8. The Lattice Problem

We have obtained the effective quasiparticle Hamiltonian for the problem to two magnetic impurities in a conduction electron sea. It is expressed in terms of even and odd states about the two impurities. By applying the set of transformations inverse to those that led to the f_e and f_0 operators it can be expressed in terms of a tight binding basis set for quasiparticles in the lattice with the two magnetic sites i and j singled out. Let us denote this by $H_{ij}(1, \ldots, N)$. We believe the essential low temperature physics of the heavy fermion lattice problem at low temperatures is given by

$$H_{eff} = \sum_{i<j} H_{ij} \tag{8.1}$$

The argument is: Suppose we solved the three-impurity problem and obtained an effective low temperature Hamiltonian H_{ijk}. H_{ijk} can be decomposed into

$$H_{ijk} = \overline{H}_{ij} + \overline{H}_{jk} + \overline{H}_{ki} + \overline{H}_{ijk} \tag{8.2}$$

where the first three terms are the parts of H_{ijk} which can be expressed in terms of pairwise interactions and the last cannot. The parts $\overline{H}_{ij}$ have the same symmetries as H_{ij} and simply renormalize the coefficients in the latter. Therefore they merely provide numerical improvement and do not add anything new. The last part consists of interactions with six (irreducible) legs. In Fermi-liquid theory they are never considered since they provide corrections to thermodynamics and transport properties that are higher powers of T/E_F than the four-legged terms in H_{ij}.

H_{eff} of eq. (8.1) therefore contain all the essential physics of the heavy fermion problem at low temperature. All Ward indentities and correlation functions and instabilities can be discussed in terms of it but not the numerical values in any given experiment. H_{eff} thus provides a Fermi-liquid theory for heavy fermions much like the Fermi-liquid theory for ^3He in which are only interrelationships among various normal state properties and indications of magnetic or superconducting instabilities are provided. To compare with experiments on actual materials, the analog of H_{eff} including the effects of orbital degeneracy, crystal fields and spin-orbit coupling must be derived or guessed.

9. Correspondence With Experiments

The theory predicts magnetic correlations among the quasiparticles at low temperatures. This is also one of the most important recent experimental discoveries [24]. The magnetic correlations have the character of itinerant electrons yet they are characteristic of interaction between quasiparticles at near-neighbor sites. This is predicted by H_{eff} and quite unlike the correlations of the pseudo-one-electron theories. In such theories, the correlations arise essentially from a Lindhard function calculated from a one-electron band structure.

The theory, of course, predicts everything can be predicted from the pseudo-one-electron approach and shows the subtleties of the problem beyond it. The density of states of the quasiparticles (qp) is specified by a modified Kondo temperature due to the magnetic correlations. The band structure of the qp's is to be obtained from a non-local potential. One of the results of underlying $<S_i \cdot S_j>$ correlations was that it is not the individual moments that are compensated; those over the length scale over which K is significant are collectively compensated. We believe the non-locality of the quasiparticle and such collective compensation are the essential ingredients for an understanding of the remarkable impurity effects observed. A (non-magnetic) impurity causes potential scattering over the region of the non-locality. The result is to change the scattering cross-section from the unitary limit πk_F^{-2} to $\sigma_{pot} \approx \pi a^2$, where a is the range of non-locality which will typically be on the scale of one or two lattice constants. To this we must add the spin-dependent cross-section of similar magnitude. This can be separated out by examining the magnetic field dependence of the impurity resistivity.

The collective compensation has the effect that even a non-magnetic impurity uncovers or deconfines the local moments around it. We believe it is for this reason that spin glass or antiferromagnetic behavior is discovered with non-magnetic impurities.

H_{eff} leads for large enough initial K to antiferromagnetic instabilities among the heavy quasiparticles. The magnetic structures are again based on simple near neighbor interactions which is remarkable for itinerant fermions. This is consistent with experiments [24].

The antiferromagnetic correlations of the qp's and the enhancement of the large momentum transfer susceptibility has important consequence in the particle-particle channel as well. One of the big puzzles in heavy fermion superconductivity has been the fact that various experimental results could only be understood [25] on the basis of an anisotropic superconducting state in which the gap function has line or lines of zeroes at the Fermi-surface. On the one hand such states are disallowed [26] in the spin triplet manifold. On the other hand, the heavy fermion phenomena is dominated by spin fluctuations and there is a strong prejudice from our experience in liquid ^{3}He that spin fluctuations [27] triplet superconductivity. This puzzle has now been resolved. Antiferromagnetic spin fluctuations promote neither conventional singlet superconductivity nor triplet superconductivity. They do promote anisotropic singlet states, all of which have lines of zeroes of the gap on the Fermi surface.

In conclusion, it seems to us that the basic principles for a qualitative understanding of all the phenomena observed in heavy fermions are now in place. For a comprehensive understanding of experimental results, phenomenological Hamiltonians of the form (8.1) including the effects of orbital degeneracy, crystal fields and spin-orbit coupling need to be studied.

Acknowledgements

The point of view expressed here is the result of collaborations with B. A. Jones, E. Abrahams, K. Miyake and S. Schmitt-Rink and discussions with G. Aeppli, P. W. Anderson, B. Batlogg, D. Bishop, Z. Fisk, H. R. Ott, and F. Steglich.

REFERENCES

1. C. M. Varma and Y. Yafet, Phys. Rev. *B13*, 2950 (1976).
2. For review of experimental results on heavy fermions, see for example G. Stewart, Rev. Mod. Phys., *56*, 755 (1984) H. R. Ott, Progress in Low Temperature Physics.
3. G. P. Meissner et al., Phys. Rev. Letters *53*, 1829 (1984).
4. For a review, see C. M. Varma, Rev. Mod. Phys. *48*, 219 (1976).
5. J. R. Schrieffer and P. Wolff, Phys. Rev. *149*, 491 (1966).
6. K. Yamada, K. Yosida, K. Hanzawa, Prog. Theor. Phys. *75*, 450 (1984).
7. C. M. Varma, Physical Rev. Letters *55*, 2723 (1985).
8. This argument is unaltered if there is a magnetic transition which involves only a part of the total local moment entropy.
9. P. Nozieres, J. Low Temp. Phys. *17*, 31 (1974).
10. C. Pethick and D. Pines (Private Communications).
11. P. W. Anderson, Comments in Solid State Physics *5*, 73 (1973).
12. K. Wilson, Rev. Mod. Phys. *47*, 773 (1975).
13. N. Andrei, K. Furuya and J. H. Lowenstein, Rev. Mod. Phys. *55*, 331 (1985).
14. A. M. Tsvelick and P. B. Wiegmann, Adv. Phys. *32*, 453 (1983).
15. K. Yamada and K. Yosida, Prog. Theor. Phys. Suppl. *46*, 244 (1970).
16. V. Zlatic and B. Horvatic, Phys. Rev. *B28*, 6904 (1983).
17. See for instance, H. Ishii, Prog. Theor. Phys. *55*, 1373 (1976).
18. The Anderson-Clogston compensation theorem is discussed by H. Shiba, Prog. Theor. Phys. *54*, 967 (1975).
19. P. Coleman, Phys. Rev. B*29*, 3035 (1984); A. Auerbach and K. Levin, Phys. Rev. Lett. *57*, 877 (1986); A. Millis and P. A. Lee, Phys. Rev. B*35*, 3394 (1987).
20. K. Miyake, T. Matsuura, H. Jichu and Y. Nagaoka, Prog. Theor. Phys. *72*, 1063 (1984); H. Jichu, T. Matsuura and Y. Kuroda, *ibid 72*, 366 (1984); H. Fukuyama, *Theory of Heavy Fermions*, eds. T. Kasuya and T. Saso (Springer Verlag, New York, 1985).
21. C. M. Varma, in *Moment Formation in Solids*, edited by W. J. Buyers, Plenum Press, New York (1984); C. M. VArma, W. Weber and L. J. Randall, Phys. Rev. B*33*, 1015 (1986); T. M. Rice and K. Ueda, Phys. Rev. Lett. *55*, 995 (1985); B. H. Brandow, Phys. Rev. B*33*, 215 (1986); P. Fazekas, preprint; H. Shiba, preprint.
22. H. Razahfimandimby, P. Fulde and J. Keller, Z. Phys. B*54*, 111 (1984).
23. B. A. Jones and C. M. Varma, Phys. Rev. Letters, *58*, 843 (1987), and to be published.
24. G. Aeppli et al., Phys. Rev. B*32*, 7597 (1985); Phys. Rev. Lett. *58*, 808 (1987).

25. S. Schmitt-Rink, K. Miyake and C. M. Varma, Phys. Rev. Lett. *57*, 2575 (1985); P. Hirschfelder, D. Vollhardt and P. Wolfle, Solid State Comm. *59*, 111 (1986).

26. G. E. Volovik and L. P. Gorkov, JETP Lett. *39*, 550 (1984); E. I. Blount, Phys. Rev. B*32*, 2935 (1985).

27. K. Miyake, S. Schmitt-Rink and C. M. Varma, Phys. Rev. *34*, 6554 (1986); D. J. Scalapino, E. Loh, Jr., and J. E. Hirsch, *ibid,* B*34*, 3480 (1986).